Personal Best:

Chasing the Wind Above and Below the Equator

Also by Edward Muesch:

Ahmed from America
Rising Above the Wave
The Land of Men

Personal Best:

Chasing the Wind Above and Below the Equator

Edward Muesch

iUniverse, Inc.
New York Bloomington Shanghai

Personal Best: Chasing the Wind Above and Below the Equator

Copyright © 2008 by Edward Muesch

All rights reserved. No part of this book may be used or reproduced by any means, graphic, electronic, or mechanical, including photocopying, recording, taping or by any information storage retrieval system without the written permission of the publisher except in the case of brief quotations embodied in critical articles and reviews.

iUniverse books may be ordered through booksellers or by contacting:

iUniverse
1663 Liberty Drive
Bloomington, IN 47403
www.iuniverse.com
1-800-Authors (1-800-288-4677)

Because of the dynamic nature of the Internet, any Web addresses or links contained in this book may have changed since publication and may no longer be valid.

The views expressed in this work are solely those of the author and do not necessarily reflect the views of the publisher, and the publisher hereby disclaims any responsibility for them.

First edition printed 2006

Second edition revised and printed 2008

ISBN: 978-0-595-50735-1 (pbk)
ISBN: 978-0-595-61629-9 (ebk)

Printed in the United States of America

This book is dedicated to my first mate and wife, Helen, who has remained nearby to weather every storm. I have crossed many oceans and continents with her always at my side.

We also wish to dedicate this book to:

The people that perished in the Tsunami.

To Prakoob Padungpol, the Thai fisherman who saved Helen's and my lives.

To Michael Fonseca and Dick York, who saved *Tahlequah*.

To the yachtsman and yachtswomen who returned to Phi Phi Don Island at great risk to themselves to help the injured and dying.

To Niracha, the nine-year-old girl who lost her father in the tsunami at Phi Phi Don Island and who represents the future of Thailand.

To each of our children and grandchildren: May they always follow their own dreams in life.

Contents

Acknowledgments ix
Prologue 1
CHAPTER 1 2001 3
CHAPTER 2 7
CHAPTER 3 15
CHAPTER 4 2002 20
CHAPTER 5 26
CHAPTER 6 32
CHAPTER 7 42
CHAPTER 8 47
CHAPTER 9 54
CHAPTER 10 2003 58
CHAPTER 11 2004 67
CHAPTER 12 74
CHAPTER 13 78
CHAPTER 14 85
CHAPTER 15 94
CHAPTER 16 102
CHAPTER 17 107
CHAPTER 18 Tsunami at Phi Phi Don Island 2004 119
CHAPTER 19 137
CHAPTER 20 144
CHAPTER 21 153

CHAPTER 22	158
CHAPTER 23	163
CHAPTER 24	167
CHAPTER 25	171
Epilogue	183

Acknowledgments

The pictures in the book were taken by Dick York, David and Sue Arnold, Stewart Milne, Jesse (*SV Gaultine II*), Michael Fonseca, and Helen and Ed Muesch.

West Marine 1500 Rally
2001

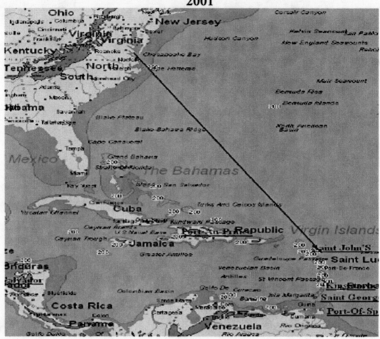

Round the World Blue Water Rally
2003—2005

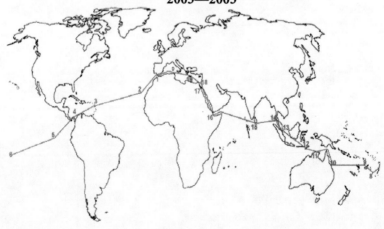

Prologue

When writing this book my editor proposed major changes to make the book more commercially desirable and in keeping with my other books. After much consideration I decided instead to write this book as a diary of events that ultimately influenced and changed my life. I wanted a book that sailor's would appreciate and identify with. In the end *Personal Best* is less about sailing than it is about not losing sight of dreams: This is our story. May our children and grandchildren be proud of us!

Sailing around the world wasn't the first time my wife and I undertook a challenge. During the 1970s, Helen and I were social activists in revolt against the establishment. Involved with the anti-war movement, Caesar Chavez Farm Workers Union, and the Quaker Project for Community Conflict, we had little time left for other things. In 1978, we left our jobs, sold our home in New York, and relocated to Western North Carolina with our children to join a Quaker Commune. For six years, we grew our own food, milked our own goats, and even butchered our own meat.

It was both a simple and difficult lifestyle, working from five in the morning until nine at night. We heated our house with wood and attempted to maintain a lifestyle in harmony with nature. Everything revolved around the Quaker Meeting, which was the center of our focus. We worked, made little, and received great satisfaction from what we did. However, as the children grew older, we realized it was time to return to corporate life to prepare for our children's future.

Our interest in sailing began many years before with the dream of some day purchasing our first boat. We couldn't afford a boat; instead, we bought binoculars and sat watching sailboats from the beach. Years later, we purchased a small 23-foot sailboat. The first time out it heeled, and we returned to the dock. Within a year, we bought another boat: a 27-foot Catalina and sailed it on Lake Lanier, Georgia. Years later, we relocated to New Jersey and joined the Toms River Yacht Club. Here we met a world-class sailor, Alicemay Wright, who helped fuel our desires to do a circumnavigation.

After 16 successful years with a major U.S. corporation, I took early retirement to move forward with our next challenge: a circumnavigation aboard a new sailboat we planned to purchase. During our voyage, we kept many personal notes and updates, which became the basis of writing our book. Helen and I wanted to leave our children and grandchildren a legacy.

Our circumnavigation aboard *Tahlequah* represented the greatest and most difficult challenge of our lives; we faced heavy weather, breakdowns, relentless schedules, and a tsunami that came close to taking our lives. I am most proud of Helen that after almost drowning she refused to give up, her only wish being to complete our circumnavigation.

Ed and Helen Muesch
S/V *Tahlequah*

CHAPTER 1
2001

Our home sold within two days of being placed on the market. We were forced to rent a house in Island Heights, New Jersey, prior to retiring. We missed the wonderful view of Toms River from our home on the cliff above the Yacht Club, where we were members during our six years living in New Jersey. Our social life revolved around the Club and our interest in Wednesday night racing. Founded in 1871, the Club boasted the oldest continuous racing trophy in the U.S. today and has several well-known members in the racing community.

Helen and I frequently felt inadequate when compared to more experienced sailors. Fortunately, we were of a mind never to be discouraged for lack of knowledge or tenacity. Our wisdom could be challenged, not our fortitude. The die was cast; we would circumnavigate.

Our first challenge was to locate and purchase a boat. We attended local boat shows and purchased equipment not knowing on what boat it would be installed. After much research, we decided upon a double ender, a Hans Christian Ketch for its sturdiness and beautiful traditional lines. Locating several Hans Christians in Annapolis, Maryland, we discovered they were in poor condition, and their owners wanted top dollar.

I received an e-mail from a woman in San Francisco who owned a 1990 Hans Christian 43 Traditional Ketch. She explained that she and her husband had purchased a boat for the purpose of doing a circumnavigation, but sadly, her husband died prior to realizing their dream. Our interest was stirred; Helen and I flew to San Francisco to see *SV Sea Child*. It was love at first sight. The boat hadn't been used. It never left the dock. The diesel engine had clocked fewer than 100 hours. Although dirty, it was in perfect condition and lacked only bright work and cleaning. Over lunch, we established what the cost would be to create a blue water yacht for long-range sailing.

After several hours of discussing and inspecting the yacht, we made a respectable offer based upon a successful sea trial and survey. Our offer was accepted, and we immediately made plans to move forward with the purchase. Two months later, *Sea Child* was ours, and she was hauled in San Francisco and prepared for the long journey east to Annapolis, Maryland. The trip took 10 days. We were standing with glasses of champagne in hand to toast her with our dear friend and inspiration, Alicemay Wright, as the truck pulled into the marina.

We had much to do to prepare her for the first season and journey to the Caribbean in November. Our first task was to rename *Sea Child*. A friend from Georgia proposed the name *Tahlequah*. During the Trail of Tears when the Cherokee Indians were marched from the southern U.S. to Oklahoma, it was agreed three chiefs would meet prior to the arriving tribes. Two chiefs arrived instead of three. One chief said to the other, "*Tahlequah*," meaning, "two is enough." Upon that spot today is the capital of the Cherokee Indian Nation, *Tahlequah*, Oklahoma. Since Helen and I planned to sail alone, the name seemed fitting. In preparation for the trip, Helen became a certified Emergency Medical Technician (EMT—First Responder) to handle routine medical emergencies. I balanced our efforts by obtaining my Ham Radio License, believing it would be useful in establishing communications and e-mail.

On weekends, our friend Alicemay Wright helped transport items to the boat from home. She stood silently watching me pass items to Helen to be stored aboard. It was obvious from the expression on her face she didn't approve of moving unnecessary items to the boat that in her mind served no other purpose than to take up valuable space. She turned a blind eye as Helen's possessions were packed below. My possessions were the straw that broke the camel's back. Passing a large bag belonging to me, Alicemay drew the line.

As the bag passed into the hold, I heard her ask, "What's that?"

Helen responded, "That's Ed's sunscreen!"

As she uttered the words, "He doesn't need all that," I saw my bag of sunscreen lotion fly through the air crashing to the ground. I waited for an opportune moment to restore my valuable bag below.

The first summer living aboard was dedicated to installing new systems and meeting safety requirements. We moved aboard *Tahlequah* in June and would leave in early November to join the West Marine 1500 for the Caribbean crossing. We had no time to sail or familiarize ourselves with the boat. Our time was occupied preparing *Tahlequah* for her first major passage of 1,200 nautical miles to the British Virgin Islands. The only pleasure we allowed ourselves was membership in the Pine Tree Naturist Resort nearby, where we relaxed around the pool on weekends. Although we worked hard, there remained time to make new friends on the cruising scene. Annapolis is an exciting town. We enjoyed showing it to Tim and Arthur (our grandchildren) and several of our Toms River friends.

I hired an electrician, (himself a sailor) to complete many of the electrical installations required, and he did an outstanding job. One day working below, he stopped, looked at me seriously, and said, "Ed, you should take one or two years to know your boat before going on a major ocean crossing."

I thought for a minute and said, "You're right. I can't dispute what you're saying. It's logical advice." His face lit up, but then I said, "I didn't retire early from a job I loved to waste two years motoring up and down the Intercoastal Waterway. We're going to the Caribbean."

Following several more attempts, he gave up and accepted my determination.

We joined the West Marine 1500 Rally because its purpose, according to its organizer Steve Black, was to help people gain blue water experience. Most Rally participants were like me, having little or no past experience voyaging long distances. Since I had only coastal cruising experience, this seemed the perfect opportunity to begin moving toward our ultimate goal of a circumnavigation.

As of that date, we had had no overnight experience, a thought always in the back of our minds. Knowing we had much to learn, what better way of learning than being in the open ocean in the company and safety of other boats? With confidence, we proceeded to prepare the boat.

The Rally had a new requirement to have a third crew person aboard if inexperienced. We immediately set about considering people we knew. A man in San Francisco whom Helen had met during sea trials came to mind. He was a live aboard, available, and appeared to be the crew person for our voyage. I wrote him a letter, requesting his experience, and got an immediate response. He was interested and had the experience needed. We set about completing our long list of safety requirements set forth by the 1500 Rally. We worked feverishly throughout the summer and achieved our goals within schedule. My good friend Bruce Kaercher at the Toms River West Marine store assisted us in tracking down parts and helping to meet schedules.

On the brisk sunny morning of September 11, I was working below on the engine while Helen was listening to music in the cockpit. Suddenly Helen's cell phone rang; it was our daughter-in-law Gretel.

"Helen, this is Gretel. I'm calling to let you know that Mark just called me. He's OK!"

"Why wouldn't our son be OK?" Helen asked.

"You haven't heard? A plane just crashed into the World Trade Center Tower. Mark works only a block and a half away."

"Thank God he's OK!" Helen said.

Without warning, the phone went dead—the result of either a weak cell phone battery or some kind of communication interruption. Suddenly, Helen heard an emergency news interruption on the radio station she was listening to. It was just announced that a second plane hit the second tower and the first tower had collapsed. When it was announced that the second tower collapsed, Helen called to me.

"Ed, you better come up here."

"What's wrong?" I asked.

"I'm not sure, but something is happening. Two planes just crashed into the World Trade Center Towers, and both have collapsed."

"What?" I asked, thinking I had misunderstood. Although we were in Annapolis, I immediately turned to look toward the direction of the New York skyline. Just then, there was another announcement that a plane had just crashed into the Pentagon building in Washington, DC. Thoughts began racing through my mind.

Are we at war? I asked myself.

Knowing that Mark worked very close to the World Trade Center Towers, I asked, "Have you heard from Mark?"

"Gretel called to say Mark was OK," Helen responded.

"Thank God."

"Gretel said Mark told her that all transportation in the city was halted. He planned to walk across the Brooklyn Bridge and try to work his way home from there."

Because the phones were overburdened with people attempting to reach family and loved ones, it was impossible to use the cell phone. Using our SAT phone, we finally managed to reach our son Mark. We hadn't realized how terrible his ordeal had been. Because he was so close to the epicenter of the Towers collapsing, the dust and debris had turned daylight into darkness. As his building was evacuated, he had to walk through storm clouds of soot and debris to leave the area.

Because Mark is a vice president of an insurance company on Wall Street and because they were one of the buildings' insurers, he was personally required to return to the World Trade Center site. Several days later while the site was still littered with bodies, this took a terrible emotional toll.

CHAPTER 2

We arrived at the City Marina in Hampton, Virginia, five days in advance of our departure date. In preparation for our arrival, the Marina was emptied to accommodate the 55 boats participating in the Rally. It was exciting to see this many boats all flying colorful burgees, signal flags, and 1500 banners, creating a carnival-like atmosphere. Large harbor tour boats passed behind our stern, announcing we were an annual group of sailboats racing to the Caribbean for the winter. Everyone seemed to stare at us in admiration as they motored by.

Every evening, the Rally sponsored parties and get-togethers in our honor, hosted by local businesses. It was an exciting time to be with other like-minded people to talk about what else: our sailing yarns. Seventy-two hours prior to our scheduled departure, a decision was made to delay due to near *perfect storm* conditions crossing our track en route to the Caribbean. Because one of the Rally participants was a weather guru, we were privileged to receive several lectures on weather forecasting during our delay.

We alerted family and friends to track our passage on the West Marine Rally website. Helen wrote home:

> *Our dream to embark on a sailing adventure is a few hours away. We leave Annapolis for Saint Michaels, Solomon's Island, Deltaville, and then to Hampton, VA, to await the beginning of the West Marine 1500 Rally on Nov. 4th. Fifty-five sailboats will participate in an informal race to Virgin Gorda, British Virgin Islands. The distance is 1200 miles taking ten to twelve days, depending on weather conditions. We have the benefit of a professional weather forecaster and a crew of three persons. We will take advantage of the radio net each day to report our position, discuss weather, and get help in solving problems if they arise.*

We left Hampton, Virginia, in November with weather experts predicting good weather and light wind. The picture quickly changed within one day of crossing the Gulf Stream. An unexpected low moved eastward causing high seas and 25-to-40-knot winds for four days. It became impossible to cook on the stove. We ate what was available without preparation. During our time at sea, it was necessary to maintain watches and sleep whenever possible, sometimes short naps in the cockpit. Not wanting to tarnish our images, I still must admit there were times each of us dealt with seasickness.

As a result of weather, several boats experienced mechanical or electrical failures, and one boat had to go to Bermuda and another to return to the U.S. Sadly, another boat became entangled in a commercial fishing net during rough weather. A French warship responded to the call. After several hours (using divers) trying to free them, the crew was removed and the sailboat set adrift. The sailboat captain refused to give the Navy permission to scuttle his boat due to being a navigational hazard. The insurance company was notified and immediately arranged for salvage. Three days later, the boat was located and successfully salvaged.

During this same period, our wind vane self-steering system broke, leaving us to hand steer the remaining 900 nautical miles. Although there were three of us to share watches (three hours on, six off), hand steering took its toll. After one hour on watch, our arms became numb attempting to fight the weather helm created by miserable conditions. Helen managed to complete her watches and maintained our course reliably.

Further south, we experienced several days of calmed winds and clear starry nights. To experience first hand how one day the seas can be mountainous and foreboding and the next serene amazed us. The end result of our experience is that we have full confidence in *Tahlequah* to handle heavy weather safely. On the seventh day, a small finch landed and hopped about *Tahlequah* for two days. He flew about the cabin, fearless we would cause him harm. We nicknamed him *Voyager*, hoping he would travel with us to Virgin Gorda, but he made the decision to leave after resting a few days. After searching the deck for *Voyager*, we accepted his decision to follow his own dreams in search of another vessel. A second memorable experience was being treated to an all-night meteor shower on a clear evening.

Our crew person neglected to mention medical problems and that he required balanced meals and medication. As the weather worsened, Helen was unable to prepare balanced meals; instead, we ate whatever could be passed up to the cockpit. Beginning to cross the Gulf Stream, I passed the helm and went below to sleep. Returning three hours later, I discovered our crew person had returned to the western edge of the Gulf Stream. He informed me it was too rough. Upset, I took the helm, once again crossing the Gulf Stream, adding six more hours to our trip.

That evening, lying below in my bunk, I saw light flashing on the deck. I asked what was happening. He responded there was a boat in the distance and wanted them to know we were here. I was concerned this could be mistaken for an emergency situation and asked him not to do this again. Learning of his medical issues, we attempted to keep him below during the day but needed to rely on him to do night watches. During the night, I checked at half-hour intervals to verify our position. Henry was an experienced sailor and under normal conditions would have been a valued crewmember. Although his medical problem interfered with his judgment, we were pleased to have his experience in several situations.

> *This is the stuff movies are made from.* Helen wrote home. *For the first few days, we thought it would have to be named PERFECT STORM II.... TODAY THE NINTH DAY is so peaceful and beautiful we have decided to name it ED AND HELEN'S GREAT ADVENTURE. Seriously, it is breathtakingly beautiful even when it is wild, and one is constantly reminded how insignificant humans really are. We are totally at the mercy of the sea and God, and it is a wonderful feeling.*

During the difficult part of our voyage, Helen stated, "I'll never do this again. Next time you go, and I'll fly there to meet you."

During the tranquil stage of our voyage, Helen seemed to forget her threat.

Reminding her of this, she answered, "It's like having a baby; you forget the pain and remember only the good part." Such is passage making, remembering the good times.

Knowing a few of our sailing friends were especially interested in making this trip next year, Helen asked, "What should we say if asked if we should we sail the 1500?"

I responded, "The answer is different for everyone. For us, it was right, and we've benefited greatly from it."

I can't say enough about the professionals who organized this event. The camaraderie and support given by fellow sailors over the radio was comforting and played a part in providing a daily safety net. Advice on dealing with various problems was shared and many problems solved. Boats rushed to the aid of others (going great distances at sea in heavy weather) where it was practical. In the end anyone crossing an ocean must understand they are responsible for themselves. Wow! Did I really say that!

Seeing landfall in the distance, we were overcome with a feeling of relief. The difficulties and hardships of our first voyage were forgotten. These were replaced by feelings of elation and the words, *WE DID IT!* For the first time, we felt like true sailors.

As we crossed the official finish line, the wind died. We started the engine only to have it sputter and stop. Repeating this process several more times with the same results, we realized we wouldn't make the awards banquet and had to anchor off until morning. The following morning, it was our crew person who discovered I hadn't switched over diesel tanks when the port side emptied—a fact I hoped he would hide from others in the Rally. The fleet took pity upon us and arranged a second party in our behalf the following evening. We were one of several boats arriving too late to attend the final awards banquet. Because we hadn't kept an accurate record of engine start and stop times, I declared this on arrival and removed *Tahlequah* from the race statistics. The following evening, we were presented with a silver tray commemorating our nautical achievement. Our favorite island drink, *Pain Killers!*, helped make up for any loss of face at being disqualified. Henry remained with us for another week. We enjoyed many beers together during that time.

Those few who complained that the Rally didn't meet their expectations were those who hadn't realized that once at sea they are responsible for themselves. What a Rally does offer is the opportunity of ongoing support (radio), including sharing weather, information concerning emergency repairs, and an alert in the event of a medical or other type of emergency. It's a good feeling knowing if I had to abandon ship there were 55 other yachts there to radio in the event of a *Mayday*. It's nice to be missed when you haven't been heard from on daily radio check-ins.

We officially arrived in the Leeward Islands, which extend from the British Virgins to Dominica. We stayed for a month in the US Virgin Islands and British Virgin Islands, visiting the many beautiful anchorages. Our plan was to be in St. Martin in January and then begin working our way south towards Trinidad for hurricane season, a requirement established by our insurer.

Last day of work—early retirement 2001

West Marine 1500 Rally—Hampton, Va. 2001

West Marine Rally from Va. To British Virgin's

SV Tahlequah on the starting line

Tahlequah's main salon

Crossing the Gulf Stream

Maho Bay, St. John—Christmas, 2001

CHAPTER 3

Even paradise requires an adjustment. We meet wonderful people, and within days, we're saying goodbye as each of us is moving on to a new place. Meeting the same people in a new anchorage brings with it a special joy and celebration. During the summer, we became good friends with Rick and BD aboard *SY Destiny*. Although they are now in Florida heading for the Bahamas, we've been able to maintain radio contact each morning.

We stayed at one location for several days awaiting a spare part from California to repair our refrigeration system. After discovering the problem wasn't the part after all, I realized I'd switched our temperature controllers from Fahrenheit to Celsius accidentally. After switching back the temperature controllers, our refrigeration functioned normally.

We began to find more remote anchorages that allowed us to enjoy the natural part of the Caribbean. We enjoyed being away from civilization for a few days and then returning for a social fix before heading out again. We came to appreciate a quiet anchorage, a place to walk on a deserted beach, bake bread, and be ourselves, swimming naked.

Helen learned to use the GPS and plotting; she was a fast learner. She also mastered the helm and the dinghy. As Helen said, shopping, cooking, cleaning all come naturally, and sometimes it's frustrating looking for a challenge. We were constantly meeting 1500 folks, and they were experiencing the same emotions we are. We had to remind ourselves this wasn't a vacation: how does one adjust to a new life style?

Homesickness is the plague of us all, especially during a holiday season, and this was our first one away from home. The telephones were particularly frustrating and the mail slow; e-mail was our precious link to sanity.

One Friday we were in Road Town and picked up a NY TIMES ... pure joy! Our friends from *SV Surge* rented a car; and we toured areas in the mountains. We had lunch at a place called Skyworld, which boasted a 360-degree view of the area. It was amazing! Road Town was our favorite place at that point in the journey. We had the best chance to mingle with local folks and shop in their department stores, bakeries, and shops.

Helen wrote home to friends:

> *We left Virgin Gorda and sailed to Road Town, Tortola, this morning. I remember when I told some of you I had gotten my EMT certification. You*

asked, "Helen, what if you are the one hurt? Well, now I can tell you.... I sustained a pretty bad gash between my left thumb and forefinger trying to grab a mooring ball. First thing I did was look at it and determine that it had to be stitched. Then I squeezed a paper towel between my fingers and proceeded to help secure the boat. Ed put the motor on the inflatable and then I told him I needed to go to the hospital.

Once ashore we requested directions to the hospital from the waterfront. We asked a man at a taxi booth, and he proceeded to give directions. A woman standing nearby overheard our conversation. Interrupting the man, she proceeded to give us different directions, saying her way was much faster. This quickly elevated into an altercation between the woman and the man, both arguing over the fastest route to the hospital as I continued bleeding. Finally, we thanked both and proceeded directly up the hill to the hospital not remembering whose directions we were following.

The hospital is not fancy but very adequate, built in 1977. After entering, the registrar said, you're in pain, get treated, and stop by on the way out. I knew I wasn't in a U.S. hospital. The ER staff was interesting.... the physician was from Nigeria, the RN from Scotland, and an aide from the UK. I needed 22 stitches. The stitches, prescription for pain medication, three visits to have bandages changed, and one to remove the stitches cost $36.53. The experience was so good that I almost didn't mind getting hurt.

Next day, we located a friend in the marina and spent the day at a Christmas market in town. She had arrived a week earlier; I told her I needed a haircut. Although there was a hairdresser at the marina, I wanted to go to the local beauty shop where she had her hair cut just for the experience. In we went and asked for Yolanda ... It was a Saturday and the shop was crowded with local children and parents.... Yolanda was a little hesitant about cutting my straight hair but did a wonderful job, and I had a great time listening to the joking and friendly atmosphere.

Christmas in the Caribbean has remained a religious holiday without the commercial trappings we're used to. The Christmas songs and music have their own Caribbean flavor and beat. People here (young and old) offer a smile, a hello, and a Merry Christmas on the street. In the evening after dinner we would go ashore and the whole town (including small children) was in the square listening to Christmas carols. There was no end to the warmth there.

After two full days in Road Town, the administrative capital of Tortola, we left for more quiet and serene anchorages, which included Norman Island. The only sign of civilization on Norman Island was a restaurant called *Billy*

Bones, and an old sailing vessel called the *Willie T* that serves as a floating pub. Norman Island is normally overcrowded with charter boats; however, on this day, we were fortunate to escape the usual crowds. The *Willie T* encourages its patrons to have more than a few drinks, go to the top deck, and jump off naked. This entitles you to purchase a T-shirt that says, *I came, I saw, I jumped.*

We befriended a pleasant couple chartering for the week and spent the evening drinking and dancing to music. Watching young naked women jumping overboard from above finally resulted in one of us saying, "If you'll do it, I'll do it."

As the evening progressed this became more and more of a challenge, aided by the rum and Cokes. Finally the four of us went to the top deck and looked over. My friend said, "I'm gonna do it."

I said, "Looks kind of far down to me."

Without further ado, he left his clothes on the deck and jumped over the rail.

His wife looked me in the eye and said, "Well!, my husband did it, aren't you?"

Hmmm, another fine predicament I'd gotten myself into. Having no choice, I left my clothes where I stood and over the rail I went to the cheers of everyone standing nearby. Given towels, we returned to the bar to drink and dance until the wee hours of the morning.

The next day we were invited to another sailboat to see the original movie *Treasure Island*. This had special meaning since the actual story is said to have been inspired by Stevenson's grandfather who visited and spent time on Norman Island. For the next month, I found myself walking about saying, "Billlllly! Booones! Where is Billy Bones?" in the same British accent I recalled from the movie. Although Helen said nothing, I suspect she was thinking, *get a life.*

Leaving Norman Island to return to Road Town each morning we discovered small fish laying about the deck. We've all heard of flying fish. I believed these to be fish jumping from the water, and the distance depended upon the imagination of the observer. Well, this observer saw a small fish jump four feet vertically and 12 feet horizontally, with its fins flapping like a bird. No I didn't have too much rum and Coke. Although told they are tasty eating by locals, we're sticking to the prepackaged variety for now.

Once Helen's hand was healed we needed to develop new skills picking up mooring balls and anchoring that no longer draw the attention of other sailboats or require stitches. To do this, we purchased two radios to communicate from the fore deck to the cockpit. Helen steers the boat toward the mooring while I grab the bridle with a boat hook. Now if the language gets lively, only we know it. Truthfully, we've had to learn to trust one another's skills and take turns sharing sailing responsibilities, which was the hard part for me.

When we returned to Road Town to have Helen's stitches removed, all went well. Helen and a woman friend from another 1500 boat arranged to take us to a gospel concert in Road Town that evening. This might sound strange to the average person, but Helen is determined to experience the local culture, make new friends, and experience a side of life we couldn't as tourists. It was a wonderful experience, people singing and dancing to Christmas music with a Caribbean flavor. I don't believe I ever saw such a sincere and spontaneous reaction; even kids were involved, and we had a wonderful time.

Each morning, we activated the Single Side Band Radio and talked with other boats. Many of the 55 boats in the Rally have scattered between the US Virgin Islands and Martinique. We continued to keep in touch every morning at 7:30 a.m. Although this net was originally set up to track our vessels from the US, many of the women crew used it to keep in contact, arrange for boats to meet at an agreed-upon anchorage, or share news.

One of the women arranged for us to get together in Wahoe Campground in Francis Bay on St. John for Christmas Day, and many of the boats were expected to be there. The plan was to remain in the British Virgin Islands until Dec. 16 and then proceed to the US Virgin Islands, spending time on St. John, St. Thomas, and St. Croix. The first week of January, we would make the overnight passage to St. Martin and stay on the French side of the island in Orient Bay (because bathing suits are not a requirement, making this our favorite place).

We arrived a few days early at Lancashire Bay, St. John, to celebrate Christmas together. For all our friends, this was the first time we would spend Christmas away from home. For the women, it was a difficult time that required adjustment. Eileen Quinn, a well-known singer of sailing songs came to entertain us after a Christmas Day luncheon in a restaurant above the campground. The setting was beautiful, and everyone was dressed in their finest. I had seen few of our friends outside T-shirts and shorts, and it was difficult recognizing anyone.

Knowing Eileen from a past Seven Seas Cruising Association gathering, Helen re-introduced herself. After talking for some time, Helen asked what songs she would sing. Mentioning the sad one about being away from grandchildren, Helen gasped saying "Oh God no! Please not that one. You can't!"

Eileen responded, "Now, Helen, you know I have to."

At the end of the evening there wasn't a dry eye in the audience. All the women were in tears.

"Thanks, Eileen!"

Helen later wrote to friends:

Truly, the adventure begins when your heart says it's OK to enjoy yourself. Christmas passed, and with January 2002, I gave myself permission to enjoy and embrace my new life.

The Green Flash is something we've heard sailors talk about for many years but passed it off as a figment of lively imaginations following end-of-day libations. We were anchored in a small isolated anchorage called Lancashire Bay located in a St. John National Park. With six other boats, we decided to do a dinghy drift just before sunset. We grew quiet watching the sun disappear on the horizon. At the final moment, all 15 of us saw the Green Flash. We can now attest to its authenticity. Captain Cook was the first to record the Green Flash in his ship's log. We mention this to discourage would-be doubters.

CHAPTER 4
2002

We remained at St. John through New Years Day, 2002. We celebrated New Year's Eve with Judy and Art aboard *SV Ciboney*. Two days later, we sailed to Virgin Gorda, where we used a mooring belonging to the Bitter End Yacht Club. We were entitled to use their facilities, which we made good use of. Our evenings were spent ashore watching movies in an outdoor pavilion for their guests. It was a wonderful place to await a suitable weather window to make our next overnight trip.

It was time to leave the British Virgins to explore new islands. Sixteen hours after leaving Virgin Gorda, we arrived at our next port of call, Marigot, St. Martin. We decided to motor-sail because the seas were confused and on the nose. We had left Virgin Gorda late afternoon for the purpose of seeing the Anegoda reefs offshore and arriving safely at daybreak. Although located in deep water between two reefs, our chart plotter showed us hard aground. We'd had this experience before; once the plotter showed us on US Highway 95, Florida. For this reason, we now check our position with several sources, including radar, paper charts, plotter, and more importantly, line of sight.

Helen and I took three-hour watches during passage. Within 50 miles of our destination, we saw the glow of lights on the horizon approaching St. Martin. Closer, we saw cruise ships and fishing boats, which we tracked on radar for safety purposes. Cruise ships rarely if ever respond to a VHF call, so we rely upon our own instincts.

We anchored in Marigot, the French capital, which looks more like the French Riviera than a small Caribbean island; here, everyone speaks French. Wine is inexpensive, and in the ships' stores, clothes are affordable. After anchoring in the harbor, we went below, collapsed for a few hours before going to Immigration to check in. Immigration took less than 15 minutes, without cost, and we were granted an automatic three-month stay. We walked the beautiful streets of Marigot, treating ourselves to a few beers and a couple of salads at a sidewalk café, one of many along the waterfront.

We searched for an ATM machine to replenish funds and found one with French instructions. Helen punched all the numbers and out came Euros instead of dollars. With disbelief, Helen held a pile of Euros in her hand and said, "What do I do with play money?" Because Euros were recently introduced, we discovered nobody wanted them, preferring U.S. dollars instead. We felt chosen to introduce the new currency to St. Martin. In stores, merchandise is priced in French francs and dollars. We focused on mastering the art of fast conversion to Euros.

On the waterfront there are always characters willing to tie your dinghy off to a cleat for one dollar or someone attempting to sell you something. One morning as I returned to the dinghy carrying jerry cans filled with diesel fuel, I was approached by a man I had seen on the dock. Realizing I wasn't going to make it alone, I accepted his help. Together, we transported the jerry cans to the boat. His name was Robert, from that day on, Robert became my helper. He seemed different from the others, never asking for more money than I offered him.

One evening, I heard someone whistling from shore and complained to Helen how annoying it was. The following day when I went to shore, Robert told me he tried to get my attention the entire night. Apparently, I had left my rental car light on. The battery of our car rental was dead.

Robert was one of those persons lost within the French system. Four years earlier, he came to St. Martin as illegal crew aboard a boat, hoping to find employment. Because he had no passport, it was impossible to obtain a work visa. Robert slept on waterfront benches, making enough money to purchase food. Without a passport, we could think of no way to help Robert escape his unfortunate situation.

In St. Martin, we prepared for the arrival of two couples visiting from the U.S. We re-provisioned at a local supermarket and awaited the arrival of our first visiting couple, good friends Grinch and Lana, a young couple in their early thirties. We met Grinch and Lana at the Pine Tree Naturist Resort in Annapolis, Maryland, the previous summer. Grinch's real name was Mark; he took the nickname of Grinch because of his colorful tattoos of the cartoon character Grinch on his chest. Mark has been our stockbroker since we became acquainted.

Following our friends' arrival, we sailed to Orient Bay, anchoring off *Club Orient* where naturists enjoy sun and sea. On the third day, we saw a sailboat anchored nearby. That evening I awoke during a dream, the boat next to us dragged in high wind during the night. I woke Helen and together we went topside. To our surprise, a neighboring boat had dragged a few hundred yards but appeared to have re-hooked itself. I returned to bed only to be awakened by Helen later informing me the neighboring sailboat was now close to the rocks. Waking Grinch, we sped off to awaken the crew of the other sailboat. We screamed, "AHOY! AHOY! AHOY *BONZA* (their boat name). WAKE UP! WAKE UP! WAKE UP!"

As we approached, a woman raised her head through the companionway. The look on her face said it all! Seems in haste to save their boat, we overlooked clothes. There we were, two naked men. She looked below and screamed, "Honey, it's the two men from San Francisco" (*Tahlequah*'s home port painted astern). Her male companion emerged from below to sum up his situation and thank us profusely. Next thing we knew they were gone; guess we'll never know if it was the fright of running aground or Grinch and me in birthday suits.

Each day we went ashore to spend the day on the beach with Grinch and Lana, having dinner at the Club Orient Papaguia Restaurant. As their time came to an end, we promised to one-day recapture the wonderful times we had together on St. Martin. After leaving, we missed their company for some time and look forward to their return in the future.

The Christmas Trade Winds finally arrived; we could now depend on daily 20-to-25-knot winds with frequent surge in anchorages. We left Club Orient and relocated to Marigot Harbor for additional protection against rolling seas. Although the anchorage was protected, we were still getting 30-knot winds (night and day) but no seas. Helen and I took turns at night checking our position on the GPS to be sure we were not dragging. The winds howl in our rigging, and it was difficult to sleep.

We discovered a local bar and hangout in Marigot, near the Marina Royalle, which we nicknamed the Star Wars Cafe. We found ourselves going there because of its uniqueness. It's a place to hang out with interesting people. Unless you were aware of its location, you would pass it by. The entrance is through a narrow tunnel with glowing psychedelic artwork on the walls. Inside you discover an open courtyard with furniture nailed to the sides of the buildings. A large tree stood in the center with furniture scattered throughout the branches to sit. Ladders accessed the furniture high off the ground. The barmaid had spiked hair, a leather spiked collar, tattoos, and leather clothes.

Here one could have a drink, breathe secondary pot smoke, have a tattoo, or have your nipples pierced while at the same time socializing. The average age of the patrons was 18–25, mostly French. Although we were more than twice their ages, we were always invited to sit at a table where some type of discussion was underway. When we attempted to sit alone, someone always came over to invite us to join their table. Looking around the courtyard, watching couples climb the tree to socialize or climb a ladder to sit on a sofa nailed to the side of the building was always entertaining. Whenever we were invited to sit at a table to talk about our liberal philosophy on life, or our early hippy days, language changed from French to English. Although these kids looked different, we always enjoyed their company.

One evening en route to the Star Wars Café from the waterfront, we took a short cut through an unlighted alley. Entering, I said to Helen, "This is stupid. We shouldn't be here, and it's dangerous." At that precise moment, a half-dozen young boys materialized from nowhere, cutting off our escape in both directions. Wielding knives and chains, they demanded money.

Responding, "I have no money" caused the situation to worsen quickly. Although I had my hand on a jack knife in my back pocket, I thought better of drawing it unless things got worse. One of them started to frisk Helen's pockets; she responded saying "You little snot-nosed kid, take your hands off me right now."

At this point, I said, "Ok, what do you want?"

"Just money," they responded.

Taking out my wallet, I said, "OK, I'll give you money, but I'm not giving you my wallet." "We don't want your wallet," one boy responded.

A boy reached into my wallet and pulled out all $300 I had in cash.

Looking him in the eye, I said, "you have to leave me something!"

Looking confused, the boy handed me back some money.

Looking at it, I said, "That's not enough; you have to give me more!"

As the boy began to give me more money, his friend smacked him in the head.

"That's enough, you fool."

Looking satisfied, the older boy said, "Look, we don't want trouble."

I said, "I don't either," adding, "let's just all walk in opposite directions."

As we walked our separate ways, the older boy stopped, turned around, and said, "I really wanna thank ya!"

The following day standing at the ATM cash machine with my debit card, I thought, *Guess I showed you guys!* Had they been older or professionals, we would likely have been injured.

Because our second visiting couple Gene and Vicki wouldn't be arriving for several weeks, we decided to visit the neighboring island of St. Kitts. In St. Kitts, we met a young German man anchored along side, sailing here from Germany in a small 30-foot boat. We invited him one evening for a beer; listening to his stories thrilled us. I quietly thought how difficult it must be to take on such a challenge alone.

On our return trip to St. Martin to meet our good friends Gene and Vicki from Princeton, we traveled to Anguilla, a British Commonwealth. The island is flat with 6,000 inhabitants and the most beautiful beaches we've seen to date. Although the island is beginning to discover limited tourism, we saw few tourists at the beaches. The beaches have fine white sand and crystal clear blue waters.

When our friends arrived, we returned to Orient Bay on St. Martin for a few days of relaxation *au naturalle*. Gene wasted no time in losing his clothes. Vicki continued wearing her bathing suit until the third day when she materialized from nowhere naked. After that, everyone was comfortable.

Gene and Vicki took us to several outstanding local French restaurants in the town of Gran Casse. On one particular day, we took a sailboat ride to the nearby island of Tintamere. At one time, this island was strictly a naturist spot for people coming across from Orient Bay. Cruise ships began to discover Tintamere Island and frequently transport guests by mega catamarans.

Helen, Vicki, Gene, and I ventured along the shore to an area known for its therapeutic lava mud. We smeared it over our bodies and allowed it to dry. Unnoticed to us, a mega catamaran had come ashore with hundreds of guests, who were now between us and our boat. Knowing we would pass them, we knew stalling wasn't the solution.

We strode along the beach through the visitors to our island. At first, no one noticed we weren't wearing clothes due to the mud, a man approached and asked if he could take our picture. It wasn't until he raised his camera to focus, he realized we were naked. We made a hasty retreat before we might appear on the cover of *Time Magazine*. Helen has a rich allover tan and my freckles continue to close gaps, but I look like I'm rusting.

Retreating to the safety of our own area, we attempted to wash off the dried mud. Unfortunately, we had waited too long, allowing the mud to harden to a cast-like state. It took an entire hour scrubbing to remove the mud.

Gene and Vicki were good sports; we hated to see them leave.

St. Martin Race Week attracted thousands of spectators and boats for the festivities associated with the event. Charter boats were everywhere, a clear threat to anyone hoping to escape damage from a dragging yacht at anchor. For this reason, we escaped to St. Barth's, a beautiful but expensive island not far away. To our dismay, we discovered beers cost $6.50 and hamburger's $12, a little much for the likes of us who had become used to good but inexpensive goodies.

St. Barth's is intended for the well to do and famous. Reputedly many Hollywood movie stars live here although we never saw one, and nobody knows where they live. A taxi driver did stop at the home of Nureyev and told us that he had discreetly picked him up at the airport to transport him back to his house where he died 11 days later. The house was simple, constructed of wood, and was located on the edge of a cliff overlooking the ocean. We sat in silence looking at the house. Because our son Paul has been in the New York Joffrey

Ballet Company, it meant all the more to us. The experience was moving, and we couldn't help but imagine how he must have loved this place.

One evening we found a place called *Key West*, a piano bar straight out of the movie *Casablanca*. A woman was singing and playing the piano. I wanted to say, *Here's looking at you kid* to anyone willing to listen. The beaches on St. Bart's are no less spectacular than any I've seen. There were beautiful lagoons, crystal clear turquoise waters, fine white sand, cliffs, and large rock formations emerging from the sea.

When we arrived in St. Bart's, we were assigned a mooring in the marina. This consisted of two mooring balls forward and aft. After helping a friend through this exercise, our time came. We attempted to repeat our first success, but it wasn't to be. We succeeded in tying to the forward mooring, but before tying to the aft, one the line snapped and wrapped around our prop. A friend jumped into our inflatable and began flying about tying lines in an effort to prevent us from drifting into another boat.

We'll call this "the red face mooring exercise." We did eventually free the prop and tie off to both moorings, thanks to our friend Ken. Our departure was easier; we didn't provide the day's entertainment for other yachts.

Having escaped the ravages of St. Martin race week, we returned to St. Martin to party with the best for the last day of the race. The races finished, Gene and Vicki returned to Princeton, New Jersey, and we found ourselves doing mundane things. When you go to the St. Martin Yacht Club and drink beer cheering as the mega yachts squeeze through the bridge entrance into Simpson Lagoon, you know it's time to move on.

CHAPTER 5

We had the opportunity to see the more remote Leeward Islands. We couldn't help but compare these to others subject to tourism. The comparison has provided a snapshot of what the islands were in past years, unspoiled and friendly—a far cry from what most have become today.

Our first port leaving St. Martin was the island of St. Eustatia. It's a Dutch island offering peace and tranquility, with old Dutch buildings, houses, and traditional Dutch villages. Holland, the mother country of Statia, is presently restoring old buildings including a 200-year-old synagogue, which was long ago vacated. Cruisers previously touring this island recommended we hire a guide. His name was Daniel, an 81-year-old man who spent the day with us. Daniel took great pride in explaining Statia was the first country to recognize the United States as a Country in 1775 by firing a 13-gun salute to the US Naval Ship, *Andrew Doria*.

President Franklin Roosevelt came to Eustatia and presented a plaque thanking the tiny island for its recognition; this plaque stands prominently in the center of the Dutch Fort today. Even more amazing, Daniel showed us a picture of himself as a boy scout witnessing the original presentation with his troop. His genuine pride and warmth brought a tear to the eye. He proudly showed us a letter he received from President Bill Clinton personally thanking him for saving the life of a drowning American woman on the shore of Statia. Viewing the spot, I have to admit the surf and undertow was such that I have no idea how he saved her. Daniel earned his emeritus award. Following the tour, we realized what made our experience so special was Daniel's love of what he did and his pleasure in sharing his life and special experiences with others. The day after Daniel showed us around his beloved island, he was flying to Holland to share the story of Statia with school children.

Everyone was friendly in Eustatia; people young and old smiled and always said hello. We found a hotel and asked if we could use the pool; "yes" was the answer, and we were provided with towels at no cost. We met a woman who moved here from Florida six years before, claiming she didn't fear being alone and never locked her doors in Eustatia.

The following day we visited St. Kitts, an Island of the British Commonwealth. St. Kitts has two main industries, tourism and sugar cane. Our guide coaxed us to stop and taste sugar cane. Sugar cane grows everywhere with train tracks circling the island for the purpose of transporting it during harvest. Old sugar plantations abound with beautifully preserved structures, many still in good condition. Most have been converted to hotels or homes. We anchored in a remote anchorage of St. Kitts to be away from the hustle of the capital city, Basseterre (meaning "low land").

Next we visited Nevis, the neighboring island of St. Kitts. Nevis and St. Kitts are considered the same country. We arrived in the capital, Charlestown, and checked in with customs. The island is also off the beaten path of cruise ships, and the people are friendly and helpful. We saw school children in uniforms everywhere, as in all the islands. Each school has its own uniform style and colors, distinguishing its students from other schools. The Nevis government provides housing and land to underprivileged residents who can't afford to purchase their own. Although the dwellings are simple, they provide shelter and basics for comfortable living. The people paint them bright island colors making them attractive.

On many islands, goats, sheep, cows and donkeys roam freely. Each town has a square with a market place where vegetables, fish, and meat are sold. We enjoyed island hopping; it's what we dreamed of and why we came to the Caribbean.

A group of us took our inflatable to shore, walked along the beach, and swam. We stopped at a local open-air beach bar for refreshment. None of us remembered to bring money. We asked the proprietor if we could pay later, all six of us.

"No problem, mon! Anything you want" was his reply.

We returned later to pay our bill and have dinner on the beach. There was an open fire using old wooden pallets and lighted torches at the tables. We had the best food we'd enjoyed in a while.

A few days later, we left for the island of Montserrat. The day after arriving at 0900, I was sitting in the cockpit drinking a cup of coffee when I smelled something burning. I turned my attention skyward to see a thick dark cloud enveloping the southern end of the island, drifting seaward for miles. I called Helen from below to observe and radioed another boat alongside. I knew the volcano on Montserrat had become active the last few days. There was the discovery of a new vent and surmised something was happening. A volcanic eruption had just occurred, and although close, it posed no serious threat to *Tahlequah*.

Most sailboats pass Montserrat without stopping. Those who stay overnight leave at first light for another destination. Two reasons for this are poor anchorage and the threat of an active volcano. Our intention was to stay overnight, rest, and leave for Guadeloupe early morning. After seeing the island, the landscape appeared like Jurassic Park with high mountains draped with clouds, sheer cliffs, and rainforests. Although this was the new capital of Montserrat replacing the one previously destroyed by the volcano, we saw few signs of life. Our thoughts after arriving were we have to remain to see this island.

In 1995 Montserrat had a population of 16,000. Since the eruption of the volcano there are less than 4,000 inhabitants. After reporting to Customs operating from a makeshift trailer, we were approached by a taxi driver by the name of George, who offered to take us on a tour of the island. Montserrat is now divided in two parts, one half being the safety zone and the remaining considered uninhabitable. No one is allowed to live in the uninhabitable zone because of the risk of more volcanic eruptions. Driving across the lush mountain landscape toward the old capital, we saw a thick cinder ash covering the road.

We saw homes, farms, churches, and businesses all abandoned in a way that suggested their occupants were away for the day. Cattle, pigs, horses, and other animals roamed freely about the countryside, abandoned by their owners. We observed a piglet that had adopted a bull, refusing to leave its side as it roamed. Cars occupied garages; doors and windows were meticulously closed; mansions, estates and common dwellings could be seen one after another.

Arriving at a mountain ridge we stood soberly looking at the capital below called Plymouth. The city was large and covered in several feet of fine white ash, looking like a scene from ground zero, Hiroshima. Thousands of homes, schools, hospitals, hotels, and businesses stood in ruins covered in white powder. There was a deafening silence; there were no birds, insects or other signs of life anywhere. The cinder ash is acidic, destroying everything it comes into contact with. In the background, we saw the volcano high above the city billowing thick black smoke skywards. We stood in silence realizing what a terrible thing this was for the thousands of people affected.

We entered a hotel overlooking the old capital seaport, now filled with ash. To enter the hotel lobby, we bent down to clear the door jam due to several feet of ash on the ground. Passing the reception desk and bar in the lobby, it was like walking through the hotel in the movie *The Shining*. We exited to the terrace and saw that the swimming pool remained full. The more we saw of Plymouth below, the more somber our mood. The lava continued to flow into the seaport causing it to expand the landmass of the island. The concrete shipping pier is now land locked and several hundred yards from the ocean.

As nightfall approached, we circled to the other side of the volcano to see glowing streams of hot lava flowing down the mountain to cover what was left of the international airport. Occasionally red glowing fireballs exploded skyward like rockets into the sky. The scene was both dramatic and foreboding and one we will never forget.

Our driver George pointed to his family's home, tearfully describing his many memories there. After leaving, we talked about how nothing had been removed from the old capital. Properties had not been pillaged for valuables, but instead remained as a testament to the people who had once lived there.

On our return drive from the capital, we stopped at a small hotel located in a temporary neutral zone. The hotel had only recently been permitted to re-open to accommodate a small group of scientists assigned to the island by the UK to monitor volcanic activity. We learned their primary function was to alert the island of impending disaster in time for mass evacuation. They seemed as surprised by our presence as we were by theirs. Following several rounds of drinks and conversation, we left to return to the new capital.

After brief visits to Guadeloupe and the Saints, we sailed south to Dominica. Dominica is located between Guadeloupe and Martinique. The contrast between the French islands and Dominica is a dramatic one. We landed in the northern part of the island, Portsmouth. Although this is a poor country, the people are friendly and helpful. Walking through the narrow streets, we were approached by vendors selling vegetables, baked goods, coffee, and other staples. Our first task was to bring laundry ashore for washing.

We motored to a dock described in the cruising guide as a place that did laundry. I waited in the dinghy while Helen took the laundry to the front door. Moments later, I saw a man run out of the woods and approach Helen. She handed him the bag and returned to the boat.

"What happened?" I asked Helen.

"The place was closed, but a man came and offered to have his sister wash it for us saying she needed the money."

Imagining the worst, I responded, "You didn't give him my silk shirts, did you?"

I was horrified and thought, *Oh well that's the last we'll see of that laundry.*

On our first trip to shore the next day, a man waved us to the beach. It was Helen's laundry man. He asked Helen for money to buy soap, as his sister didn't have any. Looking at me, Helen asked what she should do.

"We're already in pretty deep," I said, adding, "Might as well give him money to buy soap." With that, he got on a bicycle and disappeared down the road. Later when we returned from town, the same man was standing in the pouring rain riding shotgun over our dinghy. The following day, he came to our boat to deliver our laundry for more than a fair price, and the laundry was clean and neatly folded. All our experiences with the people of Dominica have been positive.

We walked through the outskirts of town passing many shacks. As it had just rained, the streets were muddy and flooded. As we stepped through puddles, the streets were crowded with adults and children playing. People smiled and said hello as we passed them. Everyone seemed to know we had come on a boat, and many said, *Hello, Captain, welcome to our country!* as they passed. At first it wasn't easy accepting that no one wanted or expected anything from us.

Dominica is a land of natural beauty, rain forests, volcanoes, overgrown jungle rivers, and beautiful waterfalls. Albert, a local guide came to our boat to offer to take us to the Indian River within the national forest. We went in a hand-built skiff that the government requires must be rowed without the assistance of a motor. Entering the river, it became narrower with mangroves overhanging the water. The trees formed a canopy that allowed little light to pass through. We could hear the parrots in the canopy above but couldn't see them. The peace and solitude in this lush land of beauty is beyond description. Albert took great pride in describing the Latin terms for the fauna and wildlife.

We heard that Dominica had a beautiful rain forest high in the mountains. Checking the sailing guide, we learned we could call a guide on ship's radio. We got Martin, another guide to respond to our call. The following day, he took us into the rain forest. We believe the reason it's called a rain forest is because it rained the entire time we were here. We stood on the edge of cliffs peering into the abyss below, all within a lush jungle.

I took pictures of Helen standing on Mangrove tree roots that towered above her. The forest was lush, and wandering from the trails could result in becoming lost. We saw old banana and pineapple plantations. Grapefruit and orange trees grew everywhere and provided us lunch. The day's highlight came when we approached an 80-foot-high waterfall. Helen jumped in and swam beneath the falls, and yes, I got a picture to prove her tenacity.

We left Dominica for our next port of call, Saint Pierre and Martinique. We could see from the water that St. Pierre was an old and beautiful village consisting of hundreds of very old traditional French dwellings along the shoreline. Above the village is an ominous volcano that erupted last in 1902, killing 29,000 persons. We saw buoys scattered within the waterfront that we later learned marked the many shipwrecks caused by the eruption. The city once had an opera house, beautiful stone dwellings, and was considered the Paris of the Caribbean. Today, there are less than 5,000 persons, and the charm of St. Pierre remains in the richness of its people and architecture.

We left St. Pierre for the nearby island of Martinique. We finally arrived in the Windward Islands that extend from Martinique to Grenada. Paul Gauguin lived and painted in Martinique around 1900 just before he left for Tahiti where he reached the pinnacle of his career. We visited a small museum dedicated to the artist and the works he painted while living here. It became clear the style of painting he is famous for began to flourish and take hold here in Martinique. We left the museum to locate a taxi, but it wasn't our good fortune to find one so we returned to the museum to request their help. Although no one spoke English, the woman at the desk closed the museum and drove us back to the city dock. It's been our experience the French reputation for sometimes being snooty is mostly unjustified.

The streets of Martinique are narrow with deep gutters lacking safety protection. One afternoon Helen and I strolled down a side street. While talking, I stumbled into a deep gutter, flying head over heels across the street and through the doors of a crowded French restaurant to the horror of the patrons. Finding myself on the floor, I picked myself up, brushed off the dirt, and proceeded to sit down at an empty table.

Seeing Helen standing at the entrance, I said, "We're here now, so we might as well have a seat and order lunch."

No one in the restaurant spoke English. Finally the waitress politely gave me a French/English dictionary. Although doing our best to order, we got something unrecognizable but made the best of the situation.

CHAPTER 6

We sailed from Dominica arriving in St. Lucia days before our son Paul, his wife Rosie, and our grandson Tim arrived for a week's vacation aboard *Tahlequah*. Once onboard, we sailed to Soufree, the National Forest and Marine Preserve in St. Lucia. As we approached Soufree and the Piton's, it looked like a scene from Asia. There were beautiful steep hills with two twin peaks dramatically emerging from the sea. I recognized the Piton Peaks from a beer bottle label the day before.

From cruising guides, we knew anchoring is prohibited so when a small boat offered to help tie us off, we were grateful for assistance. Once secured, we were besieged by kids and adults selling mangoes, bananas, bead necklaces, or willing to guide us to local points of interest. I asked the man who secured us about island tours, but refused when I discovered his price. When the marine park official came to collect the mooring ball fee, I asked again about island tours. His price was fair, and we agreed on the tour. He offered to make all the necessary arrangements; we would be picked up the following morning at our boat.

That evening, enjoying sundowners, we saw a young local man enter the water and swim toward *Tahlequah*. He brought hand-carved coconut shells, and Rosie bought two. It wasn't the last time we saw Pasqual, our uninvited guest; he continued returning to the boat bearing more gifts. He developed an interest in Rosie and used every occasion to return to *Tahlequah*; even the long swim didn't discourage him. Eager to impress Rosie, he performed back flips from our stern, after which he shared many local stories.

Next day, the island tour included a trip to the National Forest, a drive through volcano, and a waterfall we bathed in. I don't deny the National forest was beautiful, but I've become accustomed to expecting one on every island. Approaching Soufree Volcano, we smelled the strong odor of sulphur, which I identified as rotten eggs from high school chemistry. Approaching the top, I saw billowing steam and boiling black water looking like a scene from Dante's *Inferno*.

Our guide explained the history of the volcano and how it was connected to other island volcanoes. Of special interest was his explanation of how the largest crater was recently formed. Seems a local guide walked across the volcano crust to have his picture taken by tourists. Without warning, he fell through the crust and was badly burned. Today that spot has formed the largest of the crater pools. Later, we went to a botanical garden and pool heated by the volcano, and soaking in the warm water, we relaxed.

At the end of the day, our driver returned to the dock where the man who made our arrangements was waiting, visibly upset. He informed us the man who tied us off the day before accused him of stealing his business. He explained the man called his boss and threatened his life over the incident. Although I found it difficult to understand why someone I turned down would react this way, we felt bad and offered to mediate. All of us went down the street meeting several men sitting outside a house; the man I sought wasn't home. I talked with his brother explaining that the Park Ranger didn't solicit our business and that he was unaware of other arrangements. I got a simple "This doesn't concern you, mon; it's a local issue." When I tried again and got the same response, we left.

We returned to the boat and sought another anchorage along the coast. The lesson was learned: we needed to be more careful and sensitive to how local people competed and interacted with one another. A mistake we were intent on not repeating.

One morning, a local man paddled to our boat attempting to sell us bananas; that failing he asked if we had a pair of extra shoes. Our grandson responded he didn't think we had enough, I realized he was talking about me. I wore out my last pair of shoes some time ago; I wear only sandals or go bare foot. After a wonderful week with Paul, Rosie, and Tim, we returned to Rodney Bay Marina, and our family returned to the U.S.

We continued to stay in Rodney Bay Marina for another two weeks while contracting to have davits made by a local machine shop, canvas covers made for the hatches and trim, bronze polished and sealed, and finally teak trim varnished by local tradesman. During this same period, we met with old friends from New Jersey on their way to Trinidad. One of the benefits of living aboard a boat is the opportunity to meet old salts from all over the world. Ask about this place or that place, and you'll get as much detail and information as you want, sometimes more.

This morning, we added one more to our humorous incident category. Seems we mistook someone's van for a local bus. When he slowed down, four of us ran over, opened his door, jumped in, and said, "Drop us off at Rodney Bay Marina, please."

After an initial look of confusion, he responded, "Sure, mon, no problem."

Exiting the van, I gave him 2 CE's (.75 cents). It wasn't until the others attempted to pay him, he explained he had enough and was glad to help anyone. Suddenly, we realized we weren't in a public bus, and, oh well, we did it again. The day was May 2; we've been living on *Tahlequah* one year. Both Helen and I were enjoying our lifestyle and the many friends we'd made in our travels.

As we left the dock in St. Lucia, the men who installed our davits bid us farewell. They designed, fabricated, and installed the best davits I've seen on a boat. They labored for a week to complete the project in time for us to leave on schedule, taking great pride in their work.

During the installation, *Tahlequah* was tied alongside a concrete pier. During our absence, a commercial fishing boat passed, causing a wake that damaged the rail and safety stanchions. When I returned, I contracted a craftsman to make the necessary repairs. Several witnesses came forward to provide the name of the commercial boat and captain who caused the damage. Efforts to contact the captain failed, and I went to the St. Lucia Port Authority to report the incident. Within days of filing a report, they summoned the captain and collected from him the cost to repair my boat. I was impressed with their efficiency in dealing with my problem and wrote a letter expressing our thanks and appreciation.

Leaving St. Lucia, we sailed to Wallilabou Bay, St. Vincent (the Grenadines), stayed two days and left for Bequia, a nearby island. Bequia is the jewel of the Caribbean. We anchored off Princess Margaret Beach. Bequia remains a true sailing port whose culture is closely tied to whaling, model shipbuilding, sail making, marine chandleries, sail lofts, and nautical bookshops. A specialty here is scrimshaw and coral jewelry, all hand made. Two large Square Riggers are frequently at anchor here, adding to the feeling we're in another time and place.

By now my hair was a foot long, and all I was missing was the eye patch. Hmm, where did I put that thing?

We'd commissioned a model ship builder to make an exact half-scale hull model of *Tahlequah* to be displayed in the main cabin. Although Bequia continues its tradition of whaling, it must be done using traditional methods and not modern ones. There are two whaling long boats here; the harpooner uses a hand-thrown harpoon, and the methods of harvesting the whale remain as they were 100 years ago. Harpooners in recent years have been seriously injured, including the loss of limbs.

The following week, there would be a blessing of the whaling long boats, followed by whale spotters assigned as lookouts in search of Pilot Whales. Remembering we were guests in another country, we did not make judgments.

On National Heroes Day, we celebrated by visiting the other side of the island to watch model sail boat races. Both children and adults swam alongside model sailboats for the purpose of manually tacking them as they raced towards the finish line. Each village turned out to cheer their favorites in local regattas and celebrate with a community picnic afterwards. These regattas were taken as seriously as the large boat regattas and required great endurance and strength of the participants. I'd seen radio controlled model sailboat races but nothing like this. Although I admired the fortitude, I preferred changing batteries when I get tired. At the finish line, there were hundreds of people cheering favorites with music and food for everyone. We've been told this is a tradition going back 60 or more years.

One pub we frequented was named the *Whalebone Restaurant* and for good reason; the support structure and seats were made from old whalebones. The local bookshop carried hundreds of charts, nautical flags, ship's logs, and handmade rigging knives. They were many stores to provision along the waterfront.

Touring the island, we came to a whaling museum. If our taxi driver had not stopped, we would have missed it. It was a house with whalebones for handrails and fences. The museum belonged to a man who was a modern-day whaler and looked like someone out of Herman Melville's *Moby Dick*. He was a tall, lean, strong man with a thin face and piercing eyes. As his 15-month-old daughter clung to his leg, he talked slowly and distinctly, and he explained he hunted whales as did the man before him that occupied this same house. Harpoons, old pictures of whalers and paintings on whalebones hung on the wall.

We visited Moonhole, a community built in the 60s from lava rock. It's a community of 19 homes built by an American architect who envisioned a community in harmony with nature. The first home was built beneath a beautiful natural arch called Moonhole. Unfortunately, when a bolder broke loose nearly killing the inhabitants, a decision was made to build the remaining homes at the top and sides of the cliff. What makes these homes unusual is they have no windows; everything including tables, chairs, beds, etc. is constructed of lava rock, very Fred Flintstone. The community has no roads, no electricity, except for small solar panels providing minimal lighting to each home. We were reminded of our own homesteading experience back in the 1980s in the mountains of Western North Carolina.

We visited the Old Hegg Turtle Sanctuary. A man named Olden King devoted his life to saving the endangered Hawksbill Turtle, which inhabit the waters around Bequia. He gathers turtle eggs, hatches and raises them until two years old, then releases them into the ocean. His success rate for survival is 50 out of 100 as opposed to 100 in 10,000 in the wild. He showed us his large pet turtle that he walked in the surf daily. The turtle enjoyed having its shell petted and didn't want us to stop.

Once our friends Nolan and Debbie arrived from North Carolina, they adapted to life aboard *Tahlequah* and our sailing adventure like naturals. We met Nolan and Debbie a year earlier when staying at the Club Orient Naturist Resort in St. Martin. Because of all we had in common, a friendship developed between us.

We sailed to the Tobago Cays for one week to explore the islands, swim on the reefs, eat, drink, and be merry. Each day was filled with new adventures including a fish barbeque arranged for us by local fisherman on a nearby island after dark. One afternoon after snorkeling on Horse Shoe Reef, we climbed back into the dinghy. As we began to leave, a large barracuda pursued us for some distance. We were relieved when he called off the chase for more interesting pursuits. We swam and sun bathed on secluded beaches and climbed island peaks to take pictures of *Tahlequah* and other yachts below.

From anchor, we saw palm trees aflame, explosions, and clouds of black smoke rising into the sky. Closer, we saw through binoculars men fighting on the beach, evidenced by flashes we took to be gunfire. A helicopter flew overhead while large boats surrounded the island. There were swashbuckling pirates still in the Caribbean; we knew this because we'd seen them. Disney Film Studios was here in the Tobago Cays filming *Pirates of the Caribbean*. We were aware that Disney purchased Wallabou Bay, St. Vincent, to film the movie but didn't know they were also filming here in the Tobago Cays.

A large luxury yacht anchored nearby provided housing for the cast who frequently explored the area in power boats accompanied by security. We did our own thing, snorkeling on Horse Shoe Reef and exploring the small uninhabited islands near by. Helen hoped she would be recognized for a character role; I'd settle for a simple love scene. It was fascinating to observe the precision of changing locations and action every day and evening. At night, we heard explosions and saw flames in the distance; the film making process continued around the clock.

We spent a week in the Tobago Cays, enjoying the relaxation and tranquility.

If you come here without food or provisions, no problem, mon! Local fishermen are happy to provide fresh fish, lobster, coconuts, vegetables, and cook them for you on the beach if that's your thing. We've learned to break open fresh coconuts, cut the whites into thin slices using a machete, and bake them in our oven and *voila!* you have delicious coconut chips. Pour a rum and Coke, and you're on your way to another successful happy hour in paradise.

A boat boy brought us to Petit Tabogo where Disney has been filming earlier. Beautiful coral reefs with large numbers of colorful fish greeted us while snorkeling through the crystal clear blue and green waters. We saw four-eye butterfly fish, yellow goatfish, trunkfish, black durgeons, blue and yellow head wrasse, basslets, hogfish, and parrotfish. We swam holding our little plastic picture cheat cards to identify fish. Three hours later, we returned to our boat to make preparations for leaving the following day.

Late one evening, anchored in a remote and isolated anchorage, Debbie and Helen put on a lingerie fashion show for Nolan and me beneath the spreader lights on deck, and it was better than Fredrick's of Hollywood. The girls had purchased Nolan and me candy-flavored edible briefs. After the show, when I attempted to stand up, my edible briefs had melted and stuck to the deck. What could I do? Just had to be a good sport!

Sadly the time came to weigh anchor and head north to Canouan, where we swam at the resort beach. Helen bumped into the Prime Minister while doing her now famous backstroke. Nolan recognized the Prime Minister from an article he read on the airline coming from the U.S. Nolan mentioned to the Prime Minister he hadn't recognized him without clothes. That broke the ice, and for the next hour he was interested to know our views of the Grenadines as cruisers. We talked about everything from cruising fees to protecting the environment, to controlling the animal population in the islands. Dr. Ralph Gonzales was an interesting person whom we were privileged to meet.

We returned to Bequia during race week. Each day, there were parties and celebrations including music, activities, local crafts, and street stands selling local food including whale meat. Each day, we partied and had fun with other cruisers. We hiked a second time to the Turtle sanctuary, a wonderful pottery shop, and the whaling museum. We visited the home of Sam McDowell, the well-known scrimshaw artist and painter. Sam's home can only be described as an art museum located atop a cliff overlooking Friendship Bay in Bequia.

Because we were anchored in a remote area, Helen and Debbie swam topless off the boat. Discovered by a man from another boat, we found ourselves stalked daily, going out of his way to pass and wave to the women at every opportunity. It came to the point where he visited each time before going to shore. Amusing at first, we finally relocated to another area. Even this proved fruitless. Finally choosing to ignore his unwanted attention, we finally laughed it off and went about our business.

One week into whaling season, a 45-ft. bull whale was harpooned several miles off shore using traditional methods. The entire island of Bequia turned out to see the whale brought to shore. One man was responsible for cutting slab sheets of blubber to be boiled in large iron pots. A few days later, we were walking along the beach when we encountered locals barbequing whale meat. They offered us a taste and not wanting to offend anyone, we tasted a small portion.

No, it didn't taste like chicken; it tasted like beef.

The night before Nolan and Debbie left, they treated us to a special dinner and surprised us later that evening with a very special gift for Helen and me. Helen received a scrimshaw pendant, and I received a scrimshaw knife, made by Sam McDowell. We will treasure these gifts always as a reminder of the wonderful times we had together.

Nolan and Debbie sadly returned to the US as their vacation came to an end. At the airport we all hugged, and it was an emotional scene. Nolan even cried as he hugged Helen, something most men hide, and I was moved. Raised as a child in a family where emotions were hidden, this one honest act created a bond between me and Nolan. Helen cried returning from the airport, and we formed true friendships that will last many years. We look forward to a return visit and continuing our friendship and fun times together.

We received an e-mail informing us we were honored to become Commodores in the Seven Seas Cruising Association on April 1, thanks to our sponsors Alicemay Wright of *SV Ziggaret*, and Harold and Joan Robinson of *Eclipse*. We would fly our swallowtail burgee proudly and do our best to honor the clean-wake policy as we traveled around the world. We'd been Associates in the Association for 10 years and attended several of their meetings and seminars. There were now 10,000 members living aboard, many contributing to the monthly bulletins that contained valuable insights and information for cruisers on passage.

With Nolan and Debbie gone, it was now time to re-provision and return to the Tobago Cays one last time. The winds were 20 knots with small seas allowing us to safely enter the reefs. Once inside, we anchored off one of several islands surrounded by reefs, which break the ocean swells. The islands were uninhabited except for the occasional sailor on the beach, so it was easy to find a private beach on the islands complete with palm trees, white sand, and crystal clear waters. The following day, we would snorkel on the reefs, claiming a beach for ourselves.

We'd been told by our weather service the wind would increase to 30 knots with seas proportionate. We'd come to the Tobago Cays a few days early to enjoy the protection of the reefs against heavy seas and swim in the warm water until the weather improved, and we expected to be here for several days before going to Mayreau.

While in the Tobago Cays, we explored Rameau, a deserted island. We discovered a shelter complete with furnishings constructed from branches and tree limbs reminiscent of Robinson Crusoe. Although we could find no sign of life during the day, at night we saw fires burning on many of the islands. The Cays have recently become a National Park, and local inhabitants continue to camp on the islands as they have for many years. Exploring other islands, we've found similar sites and noted the people appear to maintain these islands, keeping them free of garbage, and the beaches appear to be raked.

A wooden sailing Barque similar to the one in the movie anchored nearby us. The crew spoke only French, and they would signal us every evening by blowing a conch shell. In response, we blew ours and waved, creating a bond between us. The captain wore a knife on his leg, and his mate enjoyed diving from the upper rigging; needless to mention I did not duplicate their heroics.

Helen might look good wearing a knife on her leg, I was thinking to myself.

We snorkeled on the reefs and followed schools of colorful fish living here. Although anchored without leeward protection from land, we could hear the wind howling, and the reefs protected us from large ocean swells.

We left the Tobago Cays for the island of Mayreau, where few ventured. The island had several hundred inhabitants including 46 children. There are no paved roads or electricity, giving Mayreau the distinction of the only remaining island here without these basics. The shoreline is a perfect one, white sandy beaches with palm-like shelters scattered along the waterfront. We walked up the steep hill from the beach touring the local village. The people are friendly making us feel welcome, always offering a smile and a warm hello. A few men stopped to talk, going out of their way to make us feel welcome. We discovered a small school in session and found a chair outside with a conspicuous donation box. Contributions are used to purchase pencils and paper for the children, the box says. The children of Mayreau have no opportunity to advance beyond sixth grade due to their isolation.

The anchorage had few boats; we saw large fish swimming nearby. We enjoyed swimming off *Tahlequah* before breakfast, lunch, and late afternoon. We will always remember Mayreau and the people. Before leaving, we made the commitment to collect school supplies and ship them back to the island.

We celebrated Mother's Day at the Union Island Anchorage Yacht Club by taking a room there, enjoying a candlelight dinner.

Anchoring off the Yacht Club, we saw a sign saying, *Want to have a romantic evening? Rent a Room Here for $50/night.*

Going to the front office, I stuck my head through the window and said to a very attractive young woman, "Hi, I'm here for a romantic evening."

The expression on her face said it all. Going on to explain we saw the sign and wanted to celebrate our first year at sea, she smiled. This was the first time we'd slept on land since moving aboard *Tahlequah* one year ago. Sleeping in a real bed has its advantages, not to mention a shower that's perfected the art of combining hot and cold running water. Had it not been for fear of becoming dirt dwellers, we would have extended our stay an additional day. After breakfast and a swim, we returned to *Tahlequah* becoming seafarers once again.

Although Carriacou is considered part of Granada, it's a relatively undiscovered and undeveloped Caribbean island. The previous day, we walked through the countryside and into the hills. There, we discovered many old ruins and a cemetery dating back to the early 1700s. The structures were stone and stood atop a remote hillside overlooking a beautiful lagoon. We discovered a small uninhabited dwelling atop a cliff with waves crashing against the rocks below. It was the perfect scene reminiscent of *Wuthering Heights.*

We took an interesting tour on the island of Carriacou. Unfortunately for us, it began raining as the tour began, and the van leaked badly. Helen and I sat close to avoid cascading water making its way through the window. Because there was no defroster, we couldn't see anything. Each time we arrived at an overlook, our driver offered to remain until the weather cleared, but we always declined. In spite of this, Thomas, our driver, did his best to provide us with as much information about the island as possible. Thomas knew much about herbal medicine and shared his knowledge with us; stopping at every tree and bush on the island, we received instructions on its use. I can't say we discovered a cure for cancer but did feel better when the tour was over.

We left Carriacou early morning for Grenada. Thanks to strong winds and a 10-foot following sea, we made the trip in seven hours. We passed an underwater volcano recently upgraded orange due to increased activity, called kick-em Jenny. Scientists expecting the birth of an island established a 1.5 km exclusion zone to protect us. We noticed increased current and wave activity passing nearby, which we were told was due to volcanic underwater activity. The island of Barbados was fearful of a 150-foot tidal wave that could result when the island formed.

What we've learned living on a sailboat this past year:
We see the world from a different angle, about 30 deg.
Knowing up from down is very important when
Navigating.
No matter how big a boat is it shrinks one foot per day.
Honey where did we put that thing? means I lost
it again.
Let's stay here another day, means I'm scared!
Our weather predictions are based on where
everyone else is going.
Keeping the dinghy rigid isn't suggestive.
Whalebone jewelry looks like gold
jewelry.

The length of your hair is measured in seasons
Losing your head aboard a sailboat means you've whacked it again.
God damn it! Means I broke my toe again.

CHAPTER 7

We arrived in Prickly Bay, Grenada May 17th and stayed 13 days, touring the island and making new friends. Going to shore, we passed a beautiful black hulled sailboat, crew in whites. We later learned it belonged to the Norwegian Royal family. When we passed close by to admire the sailboat, Helen waved, and the family returned the greeting. Later, Helen adapted her own impression of the Princess Margaret wave. I'm certain they believed I was paid crew. Most cruisers come together at happy hour, and this is where we meet people from around the world.

Morgan Freeman is a sailor and was having his boat, *SV Afrodesia*, put in the water while we were there. I'm told by locals he spends a lot of time sailing and is frequently seen at happy hour at the Spice Island Yacht Club, which we frequent evenings.

During our visit to Grenada, we met a Scottish couple that sailed across the Atlantic. We were amazed how much he looked like Richard Geer. Sadly, she was suffering from terminal cancer and was in much pain. One afternoon, we went with them into town to purchase morphine for her at a local drugstore. During the next two weeks, we spent many hours together, listening to music, sharing meals and beers at the Yacht Club, and an occasional tour. She had a tremendous spirit and drive that we admired. The evening before they left their boat in Grenada to fly back to Scotland, we had a final dinner together aboard *Tahlequah*. Their leaving was a poignant moment for all of us.

We took a jeep tour of Grenada, stopping to swim at a nearby waterfall. Unfortunately, two British Cricket players drowned here weeks before. The water is 40 feet deep, so swimming beneath the falls, you're forced to the bottom. After surfacing, you must go with the current, as it's impossible to swim against it. Next we went to a hot spring heated by an underground volcano. We had the spring to ourselves and lavished in its serenity and natural beauty. Sitting in the spring, we saw petroglyphs carved in the stones. We were told these were carved by the original Carib Indians who lived here until the French and British destroyed their villages and civilization early last century. Sitting here, I felt a part of their world.

A goal this season was to visit Tobago, an island off the coast of Trinidad; this would be an overnight trip. Reviewing daily weather faxes for favorable conditions, we left under ideal conditions. A few hours into the trip, instead of ideal conditions, we got 28-knot winds and sloppy seas on the nose. So much for weather projections; however, Tobago is one of the unspoiled jewels of the Caribbean and worth any effort.

During the night, we heard the bilge pump alarm, and rushing below, we heard sea water gushing into the bilge. Although the water level had almost reached the cabin sole, the bilge pump did its job, keeping us afloat. The water was coming from the stuffing box where the prop shaft enters the hull. The mechanic we hired to realign our engine forgot to replace the nuts holding the stuffing box together. I radioed the closest boat, *SV Camper Down*, asking them to standby while we attempted repairs. An hour later, we successfully repaired the stuffing box and continued to Tobago.

Arriving in Tobago exhausted, we entered the first port we came to, not an official Port of Call. Customs and Immigration took pity on us, looked the other way, and approved our entry. I'm not sure if Helen's saying, *I'm too tired, and I'm not going any damn further* had anything to do with it. I must say they were polite, understanding, and helpful. Not something I'm always accustomed to when returning through U.S. Customs and Immigration at the airport.

Although coming to Tobago to attend Race Week, we intended to meet our friends Malcolm and Margot from the *SV Kiwa* waiting for us in Englishman's Bay. Malcolm is a retired medical doctor from the UK, both avid divers (instructors), fishermen, and adventurers. Malcolm and Margot compete with one another to stay fit and ready for any new challenge. We first befriended this couple over a year ago in Carraciou and have maintained our friendship through e-mails. Although there are many cruising couples today, there are few like Malcolm and Margot who live off the sea.

I received my first spear fishing lesson from Margot today, and believe me when I say it's not easy to do. My first attempt I forgot to remove the rubber tip from the spear. Taking aim and firing, the fish gave me a dirty look as if to say, "You're kidding." My second attempt was successful, and we returned to the inflatable with our first meal. Helen prepared the fish. It was as good as any steak we've had.

Next day, we dove a vertical wall, and the jacks, groupers, parrot, and large angelfish were spectacular. The reef itself was unlike anything I'd experienced previously with every imaginable type of habitation and coral. Malcolm and I dove up to 105 feet in a marine Grand Canyon so colorful and exotic that I had to just sit on the bottom looking up the wall trying to absorb the natural beauty. Not having dived in several years, it was comforting to be with experienced divers like Malcolm and Margot.

Englishmen's Bay is a beautiful isolated anchorage that we shared with KIWA. In the mornings and evenings, we heard the Cocrico and saw Mot

Mots, and parrots. We anchored just off a reef using a stern anchor to prevent swinging. We visited a bird sanctuary and spent an afternoon in Store Bay and observed the Tobago Regatta from the shoreline. After lunch, we returned to Englishmen's Bay. After two weeks in Englishmen's Bay, it was time to return to Store Bay and check out with Immigration and Customs in Scarsborough. Because our fortress stern anchor dug itself so deeply into the sandy bottom, it was necessary to use an air bag to reclaim it. This is one island we will return to in the future.

On June 25, we left Store Bay, Tobago, for Trinidad, a 50-mile day voyage. Fifteen miles off the north shore of Trinidad, we were surrounded by a large pod of 30 dolphins that enjoyed frolicking alongside *Tahlequah*. They leaped from the water in pairs and came within touching distance of Helen and me sitting in the cockpit. Many of the dolphins had scars on their fins and torsos, an indication of encounters with fishing vessels, pleasure boats, and possibly fish nets. They remained with us for hours and seemed as entertained by us as we were by them, but finally they disappeared without warning. As the wind steadily decreased from 25 knots to one knot it was time to motor the last several miles.

Approaching Trinidad, we saw oil platforms in the distance. Entering the harbor of Chaguarmas, we saw sailboats from around the world, many freighters, ships, and barges at anchor, and a large oil platform was in the harbor undergoing repairs. We anchored in preparation to be pulled out the following day to undergo upgrades.

Going ashore, we went to a local establishment and ordered several beers and dinner costing $11.00 US for the both of us. We realized in Trinidad it cost no more to eat out then aboard. Trinidad is unique because of its mixed population; one third is Black, one third Indian, and the remainder Chinese and European. Like Singapore, political power appears to be shared. Trinidad is a large island seven miles off the coast of Venezuela and enjoys the benefits of oil, manufacturing centers for large corporations, western-style shopping malls, and many cultural activities including an Opera House and Theatre Center.

The day after arriving I jumped into the marina pool not realizing I had a pocket full of twenty-dollar Trinidad Tobago bills (TTs). Seems I dispersed them along the bottom of the pool without realizing the good fortune I had bestowed upon other swimmers. Someone hearing I lost money came to the boat and returned all of the booty. Now when I swim at the pool, there are always people nearby. Helen is threatening adult supervision until I behave.

A West Indian man, Jesse James (his real name), organizes daily trips to supermarkets, shopping malls, plays, zoos, hiking trips, and the ever-popular turtle watch. Jesse announces the day's activities on the VHF radio each morning; we have only to tell him where we wish to be picked up. Because Jesse is an asset to the cruising community, he has received two separate recognition awards from the Seven Seas Cruising Association. After our wonderful leatherback turtle-watch experience Helen wrote home:

> At 5:30 pm, we left Chaguaramas in a van with Jesse James (yes, that is his real name). He is a native of Trinidad and runs a tour service for cruisers. We drove for 1½ hours to Matura Beach. During May and June, the Leatherback turtles lay their eggs on the beach and return to the sea. There is a group of citizens of Matura Beach who have formed an organization called NATURE SEEKERS. They organize the tours in an effort to educate people about the turtles and to help protect them from poachers.
>
> We arrived at the beach at about 7:30 pm and had sandwiches and drinks that Jesse's wife had prepared for the trip. We walked the beach so our eyes could get used to the dark and we would be ready to see the turtles. Our guide led us along the water, cautioning us to be careful not to step on hatchlings that may be making their way to the water. Our tour was successful in that we saw many hatchlings and were able to witness the whole egg-laying process later on down the beach.
>
> We came across many hatchlings as you can see from the pictures. They are cute and run along the sand quickly trying to get to the sea. They look like they are swimming across the sand. We followed their trail to where the nest had been to help dig out any turtles that were having trouble. The little one in my hand was one that couldn't have made it on its own. I picked him out of the sand, and the guide told me to rub his stomach. I did, and he quickly started moving. It was such a thrill for me.
>
> Shortly after that, a guide down the beach signaled with his flashlight that a turtle had come ashore to lay her eggs. We went down. Quiet is necessary and space is left between the people and the turtle while she digs a big hole with her back flippers. After the turtle has prepared the nest (which takes about 20 minutes), she goes into a trance like state to lay 80–120 eggs. While she is in this trance, you can pet her and take pictures. When all the eggs have been laid, she spends ½ hour covering the nest and smoothing the sand in order to protect her eggs. Then she slowly returns to the sea seeming to be unaware of all the people. She will return two or three more times to lay more eggs from different sacks. This is nature's way of trying to preserve the species because the babies are in danger of birds and fish eating them before they are safe at sea.

I hope you enjoy this story and the pictures. Papa is getting tired of me saying, 'Oh if only the boys could be with us.' But he feels the same way, and we will share lots more pictures and adventures when we get home.
Love,
Helen

Helen went shopping with a woman friend, Carolyn from the *SV Spicy Lady*. Carolyn talked about her plans to circumnavigate with the Blue Water Rally in 2003. One afternoon, Helen returned informing me she told Carolyn we would participate in the Blue Water Rally.

"Is that OK?" Helen asked.

I was thrilled, as I've always wanted to do a circumnavigation, but never imagined Helen would agree. This was a dream come true.

"Yes!" was my resounding reply.

Over the following weeks, I communicated with the Blue Water Rally making final arrangements. We were now committed to begin our new adventure, joining the Rally in the San Blass Islands, Jan 4, 2004.

It was now time to put *Tahlequah* on the hard to undergo upgrades. All of this would happen within a month while we stayed at an apartment provided by the yard. Most costs here are less when compared to the U.S; a popular saying is *it's easy to go broke here saving money*. We've not had to pressure contractors; the people we employed have met commitments and maintained high standards of work. Not everyone in Trinidad has shared our experience with performance. We credit our success with paying the Yard ten percent of cost to select contractors and supervise them. In this way when there's a problem with schedule or quality, I threaten to talk with the Yard, and all problems are quickly resolved. I conclude contractors are more concerned to keep the Yard happy than me, because they depend upon the Yard for repeat business. The contractors are paid by me but through the Yard. This is a process that's worked well for us and one I wouldn't hesitate repeating in the future.

CHAPTER 8

While in Trinidad we befriended a couple, Dan and Molly from the *SV Jolly Moon*. Dan had cancer and was restricted in traveling. While living aboard their boat in Trinidad, Dan became very active with the Cocorite School—a school for disadvantaged children, mostly from broken homes. These children were self-motivated and attended school because they wanted to. Dan developed a close working relationship with the teachers and students. He organized trips for the children to visit businesses, theatres, and cultural events. Dan succeeded in getting local businesses to donate to a fund to purchase expensive steel drums for the school so they would have a band. Dan negotiated with the steel drum supplier to purchase the drums at cost.

One day Dan invited me to join him in meeting with a successful local businessman. This man had recently built a 21-million-dollar entertainment complex. During the meeting, Dan arranged for his students to visit the theater and enjoy a movie. More importantly, Dan appealed to this businessman to address the students and share with them his secret of success. He would talk about his own humble beginnings and how he built a business and became a leading financial figure in Trinidad. Dan arranged for many of these events.

Six months later, Dan was taken seriously ill, and within three months, he died. The entire student body of the Cocorite School attended Dan's memorial service. During the service, many of the students stood and talked about Dan, and how much he meant to them. I could not imagine being their age and having the courage to stand, share my inner feelings openly, and cry. They paid the greatest tribute to my friend Dan.

Having extra equipment on board I no longer needed (i.e., an anchor and a solar panel), I consigned these items to a local business. The profits from the sale of these items were to be donated to the Cocorite School in honor of Dan.

Helen has a need to help people, contributing her time and skills in any way that makes life meaningful. During our time in Trinidad, Helen met a young woman, Barbara from South Africa. Barbara had a fifteen-month-old son, Jacob. Everyone in the marina thought of Jacob as his or her grandchild. Jacob had the largest number of grandparents of any child worldwide. Barbara organized several efforts while in Trinidad. One was volunteering at a local school for disadvantaged children; another was at a local orphanage.

Each week, three women went to the orphanage to bathe and cuddle small children for the purpose of providing emotional warmth and support. Helen developed strong emotions for the children. One day Helen returned from the orphanage and said "Ed, what would you think if we adopted a small baby?"

Confused, I responded, "And you wouldn't let me have a cat!"

We both came to the realization by the time the child was entering high school, we might be entering a nursing home and made the only decision we could.

Helen began to work with a student at a local high school, helping her read and prepare for the national exams. Each Tuesday, she went to the Cocorite School to work with Kadjia, a local 14-year-old student preparing for her state examinations. School is compulsory up to sixth grade. The students must pass a state test to continue their education further. When the time came to take her exams, she passed with flying colors.

We celebrated Father's Day by hiking to Madaras Falls in the Trinidad National Rain Forest. On our return, we stopped at a local farmhouse to enjoy lunch consisting of duck, saffron rice with chic peas, salad, and home-made wine and rum.

I overheard someone make the comment, "There's not enough food."

Our guide responded, "Give them more rum."

We made good use of our hiking shoes and returned muddy, not to mention with a few blisters. It was a wonderful day, one we thoroughly enjoyed.

We've made arrangements to spend a week at Club Orient Naturist Resort in St. Martin the week of August 10. We've also made arrangements to fly to Peru and Chile to see those countries. We wanted to visit Machu Picchu (an ancient Inca city built between two mountains) and Lake Titicaca (world's highest navigable lake, 3,820 m). Because of the high altitude, we would have to acclimate ourselves to the lack of oxygen by resting 24 hours in the town of Cuzco prior to beginning our trek. Next we'd go to Chile to visit the Fjords, comparable to Alaska's inside passage where glaciers reach to sea.

This evening, we went to a David Rudder Concert in Port of Spain with other live-aboards. Life in Trinidad was a new and fun experience every day. When *Tahlequah* was put back in the water on July 1, we would stay at the Crews Inn Marina in Chaguaramas while we completed the few remaining projects. Once complete, we would sail to other ports of Trinidad, Tobago, and Venezuela to continue our voyage. Helen liked to remind me if we don't continue seeing new places and learning new things we'd die stupid, so in that spirit, it's off we go again.

Helen had an opportunity to use her EMT skills at an Internet cafe when a young woman collapsed. She refused to be taken to a hospital, so Helen stayed with her until all appeared well and someone from home came to get her. Less than two weeks later, a similar situation occurred when a man fell to the ground and was unable to get up. People attempting to help believed him to be drinking and attempted to help him stand. Fortunately, Helen intervened and learned his artificial kneecap had fractured.

The growing list of completed projects continued aboard *Tahlequah* with the installation of air conditioning and the exterior varnishing of our teak. Why air conditioning? You ask. It's because AC is the only way of avoiding mold during rainy season in Trinidad, and of course, it has nothing to do with being able to sleep nights. Below, the interior cushions were being re-upholstered and shelving installed for a new TV. The bow thruster was tested to move us in and out of the dock with ease. The new autopilot was a welcome change for long passages.

I can't say enough about life here in Trinidad; the people and culture are wonderful. We continue to see plays, dance companies, movies and lots of other activities. As for the Port of Chaguaramas, something is happening every day. There are potlucks for cruisers nightly at different marinas, organized tours to see points of interest, concerts, and everything else imaginable. Helen and I go each morning to the workout room in the marina to exercise before beginning our daily activities, followed by a swim in the pool. We recently went to the market in Port of Spain to purchase fresh fish, meat, poultry, cheese, and vegetables. There were thousands of people scurrying about buying their months supply of produce. We purchased 1½ lbs. of shrimp for $2.00 U.S., a large block of cheddar cheese for $1.50, beautiful flowers for .75 cents, and many other items for a fraction of the cost we're typically used to. I'm told these things are even cheaper in Venezuela.

Helen returned to the US to visit the kids in New Jersey and Georgia for two weeks while I remained in Trinidad supervising the many projects in progress. Our daughter Lisa returned with Helen to vacation with us for a week in Trinidad. It was an enjoyable visit during which time we re-visited our many favorite places, the turtle sanctuary, nature center, the movies, and spent time eating out and lots of time swimming in the pool.

In our ongoing effort to prepare *Tahlequah* for our upcoming adventure, we installed an electric head, electric sink in the galley, automatic fire extinguishers, and new lighting system. We rebuilt the windlass and engine to ensure reliable dependability for our upcoming circumnavigation. Helen was pleased to discover upon her return from the US her salon table refinished, new cabinetry, and spice rack. Whatever happened to the simple days of sailing? Seem the more systems we add, the more there is to fix.

Many of the islands we'll visit in the next several months are uninhabited, excluding birds, local wildlife, and the occasional sailboat. Departing Trinidad, we'll sail to Los Testigos to avoid the northern coast of Venezuela and piracy. All the courses are now entered into the navigation system, leaving us only to raise sails in the coming weeks.

On Sept. 11, we attended a special memorial service at the U.S. Embassy. Several Trinidad families who lost loved ones in the World Trade Center bombing were in attendance. It was a moving and heartfelt experience for all of us who attended. It concluded with a local chorus singing *Requiem*. After the memorial, we stopped to visit the War Memorial Museum. This was a private museum documenting many important events of World War Two. Included was the important role that Trinidad played during the war. We were unaware before visiting the museum that Trinidad housed the largest U.S. naval fleet outside the South Pacific.

Our friend Alicemay Wright flew from NJ to join us here in Trinidad. We left on Sept. 14, but had to return due to a fast-approaching weather system. After a two-day delay, we left for an overnight sail to Los Testigos, a small and charming island off the coast of Venezuela. No supplies can be purchased there as the island consists only of local fishermen and their families. We snorkeled, swam, and walked the beach, enjoying the pleasures of the cruising life.

Three days later, we sailed to Margarita, arriving mid-afternoon in the port of Porlamar. Here, we discovered we could purchase good wine for $2 a bottle or ten beers for $2 at a local sailor's hangout called Jacks. Jack is the Thai woman owner, and most cruisers can be found there at happy hour. Before leaving Jacks, we added a signed Toms River Yacht Club Burgee to the many hanging from the rafters. Needless to mention, we stocked up on wine, rum, and other necessary provisions during our stay.

Leaving Margarita, we sailed to another out island, Cubyago, where only fishermen and their families live. We went to the beach where we met a young woman with several children. Although we couldn't understand Spanish, we managed with much difficulty to communicate. She asked for a few batteries and water for her baby, which we supplied. In response, she wanted to send her husband spear fishing to provide us dinner, which we graciously refused as we had much on the boat. I took her children for a ride in the dinghy, which they enjoyed.

The following day we went to the island of Cochee, which is very poor and consists of one empty resort and a fishing village. We had a wonderful buffet lunch in the restaurant. Tame parrots flew about landing on the chairs and tables hoping for handouts. As we were finishing lunch a donkey stuck his head through the window hoping for handouts at a table next to us, but the gardener chased him.

We returned to Margarita to check out at Immigration and Customs and return to Trinidad. As the northern coast of Venezuela between Margarita and Trinidad is considered dangerous, so we joined two other boats for the return trip. The first day we anchored in Port Santos, a picturesque fishing village located in a small-protected anchorage along the northern coast.

We anchored close for purposes of security. At midnight, we were awakened by another cruiser banging on our hull informing us there were attempts to board our boats during the night, but the intruders left when high-beam spotlights were put on. A decision was made to leave at 1:00 am, head 10 miles off the coast and proceed to Trinidad. Sixteen hours later we arrived on the Trinidad island of Chacachre, just before dark.

Chacachre is an abandoned Leper Colony with many of the structures remaining including hospital, nunnery, homes, and other buildings. Due to poor holding, we dragged during a squall. The following day we sailed to the nearby island of Moons where we anchored and remained two days. Here we swam and saw many varieties of birds, including parrots, boobies, and frigate birds. Return to Trinidad we must, as Alicemay was to return to the U.S. the following day.

We completed installation of our security alarm system and had security frames installed over all hatch and companionways to prevent anyone entering. Overall our experience had been wonderful in Venezuela, the local people going out of their way to be friendly. When they asked for spare items, it was frequently their intention to offer something in return, often fish.

Having a few months left and motivated by our inability to communicate in Spanish, we decided to return to Venezuela to participate in an intensive Spanish-speaking course at the University in Merida. During this time, we lived with a Spanish-speaking family to gain more exposure to the language. The University is located in the Andes. We sat in single seats looking out the window of a small commercial passenger plane. Through the clouds were jagged mountaintops, the Andes. Dropping in altitude the plane began to turn sharply following the ice covered mountains through narrow ranges.

Although traveling extensively in many countries, I had never seen anything like this. I was preoccupied with one thought; I hoped the captain had the experience of landing here previously. Our descent ended at a six thousand-foot high plateau called the city of Merida, and towering above us was the highest peak of the Andes in Venezuela.

Twenty minutes later, we were at our destination in Merida, an apartment located in the center of the city recommended by fellow cruisers who were awaiting us there. The apartment was owned by an Italian family who immigrated to Merida following the Second World War. We enjoyed the opportunity to stay with three generations of the family in a home that their architect father designed and built. Their daughter Joya became our weekend guide, sharing the most beautiful parts of the surrounding areas with us.

Merida is a city of 100,000 people, of which 40,000 are students attending the University of the Andes. Venezuela was bordering on a civil war; we decided regardless to come here to study Spanish. Each day, we heard of large demonstrations and violence in the capital city of Caracas, but all seemed peaceful in Merida.

Our host Joya took us with our friends MaryAnn and Willis (cruising friends) to the higher altitudes, 15,000 feet. We hadn't allowed enough time to acclimate to the higher elevations; we had difficulty breathing, and any exertion left us breathless. We returned to a lower elevation in the Andes and stayed at a monastery built in 1553. The monastery was remote and offered wonderful walks on many surrounding paths and overlooks. We had a romantic dinner in the original dining room with a wonderful fire in the hearth. Our room was like it must have been in 1553; we went to bed by candlelight.

After returning to Merida, our routine consisted of waking at 6:00 am each day, enjoying a coffee pecquano, merron (small coffee with milk) before arriving at school by 8:00 am. We took two-hour classes in the morning and again in the afternoon with much homework between sessions.

The following weekend, we took the cable car to the 15,300 level, and then climbed to the top at16000 feet, this being my 57th birthday. Having had time to acclimate, we easily adjusted to the altitude. I can't describe what we saw from the ice peaks of the Andes, it's indescribable. There were several technical climbing teams both ascending and descending the slopes of the Andes.

We visited many historic sites, Old Spanish villas, mansions, and churches in the city of Merida. We befriended a young German in our class and enjoyed his company for dinner. Although school took much effort and study, we enjoyed daily excursions throughout the town, with an opportunity to practice our Spanish. Although Helen has far superior language skills than I, it didn't prevent me from muttering whatever words I could remember from the day's lesson.

The city of Merida reminds me of the Spain I visited over 30 years ago. Contrary to all the rumors we heard prior to arriving, the streets were safe to walk. The people were warm, friendly, and helpful. The final day of classes our instructor asked Helen to provide a birthday cake to celebrate my 57th birthday with the class. Everyone sang happy birthday in Spanish, and I will remember and cherish this birthday always.

CHAPTER 9

We awoke to a loud knocking on the hull; the time was 1:05 am. Scurrying to respond, I removed the hatch boards to be told by a hotel employee we had an important phone call, and we needed to come to the lobby right away. We knew something was wrong, but what? We knew our son Ian had died in a motorcycle accident two years earlier, leaving a wife that was severely depressed. We could only imagine a tragedy occurred.

Arriving in the lobby, I accepted the phone from the desk attendant.

"Hello," I said.

"Ed, this is Mark. Are you sitting down?"

"What's wrong, Mark?" I replied.

"Jesse was killed in a motorcycle accident this evening." I hadn't heard him right; my son couldn't be dead. It was impossible.

"Ed, did you hear me?" Mark repeated three times.

"I have to go Mark; I'll talk to you later," was all I could say.

Putting the phone down, Helen looked at me, "What's wrong, Ed? Is it Sharon?"

Saying nothing I walked with Helen through the hotel door and outside.

Putting my arms around Helen, I began shaking and burst into tears. It took several minutes for me to get out the words, the most difficult words I'd ever spoken.

"Jesse was killed this afternoon," I said.

Helen appeared in shock, and she said nothing. We stood there holding one another in disbelief. Several minutes later Helen said faintly, "We have to go home."

Walking back into the hotel lobby, I handed the woman at the desk a piece of paper asking her to call our son Jesse's house. Instead, she called Jesse James, the Trinidad man who cared for all Cruisers in Trinidad.

Handing me the phone, Jesse said. "Ed, I'm so sorry, I'll handle everything for you."

"We need to return home in the morning." I said.

Replying, Jesse said, "I'll meet you at Crews Inn at 8:00 am. I'll have your plane tickets, and I'll take care of immigration and customs. I'll explain everything to them and take you to the airport."

I lay in bed that night dreaming this never happened. Somehow I had misunderstood Mark, and all of this was a bad dream. My world was devastated, and I had nothing left. Our son Jesse, the child that Helen and I created together, the child that bonded two families, the child that represented our dreams was gone. Jesse's college sweetheart, now his wife of one year was at home alone, and we could do nothing for her. Paul, our son, and his wife Rosie would help Brigitte make whatever arrangements were necessary until we arrived.

Jesse James, our Trinidian friend, took us to the airport in the morning. He contacted the airport and ticket counter before our arrival. We were expedited from the back of a long line to the front for quick processing. We were upgraded to the front of the plane for comfort; all arranged by Jesse James, the true friend of all cruisers. Sitting silently in the plane looking out the window, my childhood memories of Jesse came cascading back.

How could this happen? I kept asking myself. Why? It makes no sense. Jesse was just completing graduate school and would be finished in a few weeks; he had everything to live for. He had a beautiful and wonderful wife, a good job, and a home. I had no answers for this senseless tragedy, only questions.

Arriving later that evening in Atlanta, Georgia, our son Paul picked us up at the airport and took us directly to Jesse and Brigitte's home where everyone was waiting. Brigitte's family and all our children stood in the driveway, hugging one another. The next few days were the most difficult of our lives. Brigitte's family and ours came together to support each other during this most difficult of times.

At the memorial service arranged by our son Paul, hundreds of Jesse and Brigitte's friends came to pay respects and bid our son Jesse farewell. Jesse was an athlete in high school playing discus, football, and basketball. His coaches, teachers, and friends from high school and college came to pay their last respects. Tearfully, I talked of our son Jesse and what an important part of our lives he was.

Since losing our son Jesse, life doesn't have the same glitter. Jesse was the one person in the family everyone looked up to and admired. Jesse called everyone in the family routinely, visited more often than anyone, and could always be relied upon to show up or call when least expected. *The Atlanta Constitution* newspaper once referred to our son in their athletics section as *The Gentle Giant*. We will always think of our son Jesse that way.

Losing a child is the most difficult part of life; no one should have to experience it. Jesse lives on in our hearts and will be with us always. Helen reminds me having Jesse was a gift and to cherish the wonderful memories we have of him. Some days our loss is so great we cry inside and must remind ourselves of the wonderful times we shared with Jesse as a child and an adult.

I don't need to remember our son Jesse as a good athlete, a good student. Instead I need only to remember how proud he made me feel inside. I will always remember Jesse as a person who never said a bad word about anyone and how I always looked up to him because of that. I remembered how Jesse's high school South Eastern discus record remains unbroken and how he never gloated when achieving it. Jesse's coach in high school once told me that *Jesse was a true champion*!

One of the most important parts of our life has been our extended family. We began this voyage as an example for them, to follow their own dreams in life. There is no greater tragedy in life than for a young person to die. Losing two sons, Ian and Jesse caused us to question our own purpose and being in life. It's because of this we decided to put our thoughts and feelings in a chapter dedicated for this purpose. Anything we do in life is influenced by our loss. Following the funeral and because arrangements were previously made we needed to spend time alone at the beach in an effort to make sense of what had happened.

We returned to Trinidad in preparation for the arrival of Brigitte, Jesse's wife. She was to join us here in Trinidad aboard *Tahlequah*. Jesse and Brigitte were supposed to come to Trinidad together, Brigitte wanted to fulfill their promise. She arrived; we did many fun things together while here. Shortly we begin our trek west to the many beautiful out islands, the ABC's and the San Blass Islands where we'll officially join the Blue Water Rally.

We cancelled our trip to Peru and Chile. With the loss of our son Jesse, our hearts were no longer in it. A month later, Helen wrote home:

To Family and Friends,

Once again, we cannot thank those of you enough who wrote to express your sympathy and support during this difficult time. These past weeks have been the saddest and most difficult time of our lives with the loss of our son Jesse.

Our hearts go out to his wife Brigitte, her parents Mel and Martine and the rest of our family who are very close. Our family has pulled together to support one another through a second tragic loss in only two years. Although we've returned to live aboard Tahlequah here in Trinidad nothing seems to be the same.

Love,
Helen

A few weeks later, Helen wrote home again:

To Family and Friends,

Time moves on but not our hearts, even after months, our hearts are broken. It's been a while since our last update; it remains difficult to return to where we were prior to losing our son Jesse. The good news is that Brigitte, Jesse's wife will fly to Trinidad to spend the Thanksgiving holidays with us here in Trinidad. Our son Mark flew to Trinidad on business, and we enjoyed his company aboard Tahlequah during the weekend he was here. We saw a local championship cricket match with the local team winning. We continue to see plays, concerts, and the World Pan competition held here in Trinidad.
Love,
Helen

CHAPTER 10
2003

Although losing our beloved son Jesse, we felt he would want us to continue our voyage. We had mixed emotions leaving Trinidad, a home away from home. Having delayed our departure for engine rebuilding, it was now time to begin our circumnavigation. We planned to work our way from Trinidad towards the San Blass Islands, where we officially joined the Blue Water Rally. The San Blass Islands are remote islands where the indigenous Kuna Indians continue to live unchanged as they have for hundreds of years. We hoped to trade or purchase Molas (hand sown colorful designs) from the Kuna Islands.

We were fortunate to have the company of another sailboat, *SV Second Wind*, traveling with us from Trinidad through Bonaire. Leaving Trinidad late afternoon, we arrived at the island of Los Testigos, at sunrise the next morning. We stopped here for only one day before sailing to Margarita. The sail was one of the most beautiful we enjoyed. Pods of dolphins swam aside our hull leaping from the water as we arrived in Margarita late afternoon. Although planning to stay at the Hilton Marina, a last-minute decision was made to stay in the Porlamar Anchorage and enjoy the many local seaside restaurants and services there.

Margarita is a duty-free port established for the Venezuela wealthy to shop and vacation. Modern malls, stores, and supermarkets abound. Due to the economy, dollars are worth more on the street than the official exchange rate. Beers cost .30 cents each; dinner costs less to eat out than on the boat. Although Margarita is considered inexpensive by westerners, the Venezuelan economy is in a state of chaos benefiting no one. We prefer goods and services to be fairly priced for the benefit of everyone.

We previously bought an apartment in Margarita in a high-rise building located on the water; our plan was to return here to spend the Christmas Holidays. When we completed our circumnavigation, our long-term plan was to spend summers in the U.S. and winters here in Margarita.

After nine days in Margarita, we left for Tortugas, a Venezuelan island 90 miles west of Margarita. Tortugas is a low-lying island with beautiful white sandy beaches and crystal clear waters. Local fishing boats come here, and a few fishermen live in a small fishing village on the island. We picnicked at a deserted beach with our friends Marty and Michelle from *Second Wind*. The following day we began another overnight sail to the Los Roques, a Venezuelan National

Park made up of many small islands. We paid special attention to our position, not to become one of the many statistics lying scattered among the reefs.

In the Roques, we met Bill and Ruth of *SV Sea Bride*, our good New Zealand friends whom we met one year before in the Grenadines. Entering the anchorage, Bill and Ruth were dressed in their native Maori dress and began the Harka, the Maori War Dance. Ruth sported a chin tattoo that Maori women were once known for and her traditional Maori dress and beads. We roared with laughter, giving them a big cheer as they performed their dance; it was a wonderful greeting after an all night sail. We anchored off the island of Sarqui and went ashore; small crabs and lizards took an interest in our occupying their beach. Large groups of pelicans flew overhead.

Rushing to maintain a schedule, we next sailed one day to the Aves, another small group of islands off the Venezuelan coast. The Aves are two separate island archipelagos, separated by 10 miles of deep water. They got their name from large colonies of Boobie birds living there. A large horseshoe reef gives protection to the islands smaller cays. We entered a reef-strewn lagoon with thousands of Booby Birds circling above the trees and nesting with their young. The young Boobie Birds were large, white, and fluffy and sat in the trees waiting to be fed. At first they appeared to be large blooms on tree branches. Two palm trees stood high above the white sandy island, appearing like aliens under a full moon beckoning us to the beach.

Two days later, driven by schedule, we departed from the western Aves. Following a seven-hour sail, we arrived in Bonaire, a diver's paradise. The waters here are crystal clear, the downtown area and waterfront filled with colorful shops, restaurants, and buildings brightly painted. Because this is a tourist-based economy, everything here is expensive compared to other islands. The legal tender is the Dutch Gilder, but everyone accepts dollars.

We took a taxi ride of the island and were surprised to discover the terrain is like Arizona, arid and desert-like with cacti everywhere. Colorful restaurants and waterside businesses abound. We stayed at a marina in Bonaire that was less than a 30-minute walk into town. We walked frequently, enjoying the many interesting sites along the waterfront. We saw wild donkeys, many of which have now been placed in a donkey sanctuary created through private donations.

Two weeks later, we left for Curacao, another beautiful Dutch island. The main harbor, Willemstad, is a miniature copy of Amsterdam. The colorful European buildings line the waterfront, the pontoon bridge makes for a wonderful tourist photo opportunity.

The harbor we're told is the third-largest harbor in the world, and it's a major oil depot with tankers and cruise ships passing in an endless procession along the waterfront. Here we enjoyed lunch at an old café built in the late 1700s. Everyone speaks Dutch or a derivative of Dutch called Papiamento. This time of year, Hollanders come to Curacao to vacation and escape the cold winter of their country.

No story about Curacao would be complete without including the story I e-mailed home shortly after arriving there; it's a true story.

Dear Friends,

Those of us who enjoy vintage wines and are old enough to recall the events of the years they were bottled might recall Ian Fleming's blockbuster film, DR. NO (a mad scientist intent on taking over the world from his hideaway on a mysterious island).

As cruisers, we encounter people and places we might otherwise never come into contact with in the normal existence of every day life. We've become accustomed to being considered a nuisance to immigration and customs officials although they remain polite and helpful. I was once told cruisers are considered the dregs of society, living on derelict boats and polluting otherwise beautiful waters. We think of ourselves as adventurers and protectors of the environment; the truth lies somewhere in between. Now for our story:

DR NO'S MYSTERIOUS ISLAND

In appreciation for the wonderful service provided by our benefactor, I won't mention the specific location or identify him by name because of his need for privacy. Our story began when a friend recommended we stay here while waiting to join the Rally. Although there were no advertisements in the guides, I managed to acquire the phone number of this small privately owned and secluded marina. Assured by the Dock Master he would find room for us upon arrival, we sailed to our new destination. The small marina is located deep inside an inlet, as we approached; a military style security boat circled us with three armed guards. Two men on the dock motioned to us to follow their instructions as the security boat escorted us in.

Once secure we went to the office to make arrangements for our stay. Our first observation was the security boats scattered throughout the marina, armed guards patrolling by foot, and four wheel drive vehicles. The location is isolated and surreal because of the Monument Valley like foothills comprising the landscape. Within the marina, no other sign of nearby life was visible. I saw what appeared to be a secondary road well guarded within the complex that went to another secluded area of the complex. More armed guards were stationed to the entrance of this road. The word was out that no one was permitted to venture into this restricted area.

Leaving the office, I mentioned to Helen the monthly fee in my opinion didn't seem to cover the cost of our staying here. Although the marina was a small one there seemed to be no one there except for a few couples like ourselves making us feel special. Although this place was a large complex, it lacked any amenities normally associated with a resort or marina, no hotel, restaurant, golf course, store to purchase supplies, etc.

There was a beautiful beach open to the public on weekends for a nominal charge, but otherwise there was nothing but a few thousand acres of arid desert land with beautiful foothills reminiscent of my childhood in Arizona. Security boats with armed guards patrolled the waters off the beach seemingly watching everyone. Security guards provided us with transportation to the beach or the front gate if we wished at no charge.

After exploring the area for a few weeks, we noticed there was an inner ring of security that protected an area reportedly occupied by two homes, invisible from the road and water. Who lived there was the question as the security certainly wasn't for us in the marina. One afternoon, Helen drove to the beach to discover there was no one there except security, and all the beautiful furniture had been removed. We were informed by security the beach was no longer open to the public.

We as residents of the marina were welcome to use it anytime we wished. Security had increased since we were away, we were carefully screened at the gate, and a phone call was made to marina security alerting them we were coming. A security card provided entry to marina parking. All night guards patrolled the docks and unlighted high-speed security boats patrolled the waters around the island including our marina.

Although we were often the only people at the beach, there remained security guards patrolling the sands. I'm happy to say that now the beach is closed to outside visitors, our benefactor enjoys coming to the beach for an afternoon swim. We've since learned in the future the marina will be converted for his private use, but in the meantime we're considered special guests who can enjoy our stay and security here. Although some of us might consider this strange, we wish to thank Dr. No for his generosity and our appreciation for being special guests.

We understand that many wealthy people in this part of the world are targeted for kidnapping, ransom, and extortion and are forced to take reasonable precautions for their safety.

Ed and Helen
SV Tahlequah

Other guests we met there were Ed Atkin and Bernie Houston of the *SV Oriani*. Ed and Bernie authored the book *One Wave at a Time*, detailing their 20-year adventure circumnavigating. We've been fortunate to share times together hearing about their many adventures throughout the years. We purchased a signed copy of their book; above all, we thoroughly enjoyed reading it. Knowing that one day we would write our own book, they were an inspiration to us.

We toured the Anne Frank Museum, a copy of the Anne Frank Annex in Holland. One of the most extensive slave museums in the world can be found here, as well as a 400-year-old synagogue still used today. We visited the synagogue for Saturday services and were welcomed and treated as special guests. A man sitting close by helped us understand the service. The synagogue is a duplicate of the one in Amsterdam, and has sand-covered floors.

Leaving *Tahlequah* in the capable hands of our trusted friends Marty and Michelle (*S/V Second Wind*), we returned to the U.S. for our annual pilgrimage home. This year was different from past years in that we returned to the U.S. to visit family and friends for Thanksgiving holidays instead of the Christmas holidays. Like past years, this has again proven to be a time of centering down and regrouping ourselves.

Mark (our son) surprised us by arranging a dinner at the New York Yacht Club to celebrate my birthday. I wasn't above pocketing a few paper napkins from the men's room to impress my sailing friends back in the Caribbean. It was a wonderful evening and one I won't forget. Mark promises to take us back next year if I don't pilfer more paper napkins.

Returning from the U.S. with so many boat items we've agreed to the self-imposed rule of disposing of equal weight to maintain our water line. Otherwise, as one friend suggested, the bottom paint will have to be raised to the rail. After returning to Curacao our friends Marty and Michelle left on their own journey to Cancun, Mexico. After having sailed four months together, it was a bittersweet departure. Bitter because we hate to see them leave but sweet because as sailors we follow our dreams wherever they take us.

Now mid-December, we flew to our apartment in Margarita to enjoy Christmas here. Although here only for a short period this year, we were looking forward to making the most of our time here. In other words, the beach, the pool, and eating out: the three basic fundamentals of life when you are retired. This brief period was important to us before beginning our circumnavigation.

Arriving in Margarita, we immediately set about purchasing the many things we needed for our new apartment: TV, cooking utensils, wall hangings, and decorations. The apartment was truly beginning to feel like home. Each evening, we enjoyed a cigar and a rum and Coke, as we took in the beautiful ocean view from our sixth-floor outside patio. Each day, we swam in the pool and walked the beach. Occasionally, we went to Jacks restaurant for drinks.

In Margarita, we befriended an Englishman, a retired engineer living in Margarita. Although he was 70 years old, he was married to a wonderful Venezuelan woman in her early thirties. They had two young children, a boy and a girl that Chris adored. Chris looked after our home when we were away, he paid our bills, saw that the condo was cleaned, and arranged for maintenance when necessary. We enjoyed having Chris and his family visit when we were in town.

This year, we received an invitation to come to his home for Christmas Eve. Being away from our own family through the holidays made this invitation special. There were many relatives including small children and teenagers. We spoke very little Spanish, and no one spoke English except for Chris' wife. Being in a home with small children helped us to forget how much we missed our own grandchildren. It was a Christmas we will always remember.

We flew to a naturist resort on Bonaire to spend a week relaxing and soaking up the sun. The resort is located on the far end of the island next to a sailboard school. As usual, when we go to a naturist resort, we make new friends that enhance our experience. We've made many close friends this way over the years. Following dinner, we saw the New Year in, celebrating it Dutch style with wonderful drinks, candies, cakes, and music.

The destroyed capital of Montseratt—Volcano erupting in background

Petit St. Nevis

Bequia Whaling Museum

Trinidad—Jesse's wife Brigitte & Helen

Merida, Venezuella—top of the Andes

Sorobon, Bonaire—An early morning walk on the beach.

CHAPTER 11
2004

Returning to Curacao, our good friends Ivor and Bernice aboard *Safari* came to join us. Together, we made preparations during the coming weeks for our departure for the San Blas Islands and our adventure with the Blue Water Rally. The trip from Curacao to the San Blas Islands is considered the fifth most difficult worldwide leg, because of unpredictable weather and seas. We will carefully select a good weather window with an opportunity of ducking into Cartagena if necessary.

We had yet to sail a leg without some type of breakdown, the trip from Curacao to the San Blas Islands was no different. Within 12 hours of leaving, steep following sea began to build from astern. The base plate supporting the autopilot broke loose, requiring temporary repair. At the same time, the packing gland in the stuffing box was leaking and required adjustment. While Helen took the helm, I went below to do what could be done. I rarely get seasick; however, this was the exception. The heat below combined with the smell of diesel fuel and excessive rolling caused me to get ill. Three hours later, the repairs were made, and we returned to using the autopilot.

Five days and 700 miles after leaving Curacao we arrived in the San Blas Islands off the Panama coast. We smelled burning wood from 20 miles offshore. We delayed our entry until daybreak so Helen could stand on the pulpit watching for reefs ahead. Following the coordinates provided by the Blue Water Rally, we anchored off the island of Iskadup. The islands are inhabited by the Kuna Indians who have chosen not to give up their traditional lifestyle. Their success in maintaining their culture is due to their isolation and freedom from tourism, excepting sailboats.

Their villages are constructed with thatched roofs and walls supported by sticks. The dirt streets are filled with young children who follow us wanting to hold hands; no one begs or asks for anything. Our friend Ivor offered three children one soda; each took one sip and continued to pass it amongst the others until it was gone.

The women dress in traditional clothing, having beautiful hand-sewn Kuna patterns called Molas. Many have nose rings and tattoos; they are adorned with beautiful beads, which they wrap around their arms and legs for decoration. Their mode of transportation continues to be hand-paddled canoes made from trees; young children paddle to our boats. Helen offered one canoe of children lollipops, and soon every child in the village came to *Tahlequah* for candy.

Only children came to our boats as the village chief forbade villagers from bothering visiting boats. On one occasion, the village chief allowed the women to present their molas to us for purchase when we visited their village. Molas are hand sewn, intricate and relatively inexpensive. Before leaving the San Blas Islands, Helen bought many molas from mola makers who paddled out to our anchorages, sometimes far away from their village. We could only imagine how long it would take them to reach us. There was never pressure, and "no" was cheerfully accepted by the Kuna Indians.

On rare occasion, a young male is selected by the village to become a master molar maker. He is dressed and raised as a girl through adolescence. As a child, he receives special training by the women of the village to become a master molar maker. While anchored far from the village we were visited by a master molar maker. Arriving in a dugout canoe, one man paddled while another held a shade umbrella over the master.

Although the Kuna Indians are poor, we saw he was wearing gold rings and jewelry. Having all the mannerisms of a female, he laid out one mola at a time for us to inspect. He explained in detail each mola design and how it was made. His molas cost between fifteen to forty dollars, and were exquisite. Molas made by the women of the village cost between five to 20 dollars. After purchasing several molas, we realized we had enough to open a mola shop in the U.S.

One afternoon, an elderly man paddled to *Tahlequah* and appealed for help with their school. We donated pens, pads, and eyeglasses provided to us for that purpose by our dear friend, Alicemay, who visited there in 1985. There are many albino children who need sunglasses due to their sensitivity to light. Because Alicemay knew this, we were able to donate sunglasses she provided for them.

One evening the Blue Water Rally arranged for the village to provide a dinner of fish and rice, entertaining us with traditional dance and music. The men played flutes as they danced around the women. The steps were unlike anything I'd seen and the energy level extremely high requiring great skill. In the final hour, cruisers were invited to dance, some did so with varying levels of skill. Yours truly sat, ate, and watched.

We left the San Blas Islands with Safari sailing to Portobello, north of Colon. There Rally Boats regrouped, enjoying a special luncheon at a local restaurant overlooking the port. Portobello, once a major port in Panama today, is a small picturesque town with many Spanish type structures including a beautiful fort.

A couple we met in Annapolis crewed for us during this period. Their contribution to watches was greatly appreciated.

After two days in Portobello, we left for a day sail to Colon, entrance to the Panama Canal. On route we picked up many large vessels on radar. Approaching we were given permission to enter the Break Water by the Panama Canal Signal Station. Fearing the number of ships and their intentions, we attempted to reach one cargo carrier by radio without success.

Coming within one mile of the vessel we made a major course correction, suspiciously the vessel seemed to swing once again in our direction. We put out more radio calls, again we received no reply. Pointing *Tahlequah* towards land the ship once again swung in our direction. I radioed a final call stating our position, and intentions. In desperation I made a 180 deg. course change attempting to escape the goliath vessel. It was only then we realized the vessel was anchored outside the Panama Canal Breakwater awaiting clearance to enter.

Seems the wind caused the boat to swing in our direction each time we altered course making it appear they were attempting to engage us. Next radar I purchase will have multiple targets tracking capability.

Entering the breakwater we were given permission to anchor in the Flats outside the Panama Canal Yacht Club. Here we gathered with other Rally Yachts and were given briefings on the Canal Transit and procedure. Each evening we drank and ate at the Panama Yacht Club. The following day we were taken in vans to the many government agencies to complete the necessary paperwork required to transit the canal.

The Rally successfully negotiated our transit with a flotilla of three yachts in dedicated locks to ensure a safe transit. Because we already had three crew on board it was necessary for us to provide only one additional line handler. Rather than accept a Rally volunteer we elected to hire a professional line handler whose name was Alex. Alex was a professional line handler for many years, all the pilots knew him well.

Prior to transit we were required to have an official measurement of *Tahlequah*. A pilot boat dropped a measurer off at the appointed time. Walking topsides with his combat boots and tape measure we discovered he was tracking gum all over our deck. Apologizing he removed his shoes and went about taking measurements. He measured from the farthest point forward to the farthest point stern. Seeing the dinghy missing he wanted to know its width for the purpose of adding it to the davits. We went from a forty-three foot yacht to a fifty-six foot yacht when considering the davits and bowsprit. Fortunately for us the cost of our transit was included in the Rally fee.

On Feb. 18th at 0600 a Pilot boarded our vessel and instructed we weigh anchor and proceed to the first of six locks referred to as the Gatun Locks. It was decided due to our size we would be the center boat providing power for the other two boats secured alongside. A large container ship entered the lock in front of us. With instructions from the pilot I powered the three sailboats into the lock attempting to keep us centered.

Canal Line Handlers from atop the walls threw monkey fists to our line handlers aboard the outside boats. The boat line handlers fastened the two outside lines, which were then raised and fastened at the top of the walls. Constant adjustments were made aboard each yacht to keep us centered in the lock as the water filled raising us to the top of the lock. We proceeded to the second and third lock in the same manner. There were observation towers with tourists waving from above. I heard a voice calling my name, looking up I saw Peter (organizer of the Blue Water Rally) taking our picture as we waved back.

After the third and final lock we broke down the flotilla before entering the Gatun Lake. Our pilot Alex allowed us to motor sail twenty-eight miles to the next lock. Here we passed through the lock dedicated to only Rally Yachts. We motored to the third location of the two final locks that connect to the Pacific. Because the final lock is the most difficult of the six locks we had some trepidation about what to expect.

Due to the expertise of our Pilot there were no mishaps and our exit from the final lock and entrance into the Pacific went without incident. Once out of the lock we broke down the flotilla, each of us was now free to motor to the Balboa Yacht Club where moorings were assigned.

Quickly the Bridge of the Americas came into view officially signaling our welcome to the Pacific. Our Pilot informed the crew since it was my first time as skipper through the canal it was their responsibility to throw a bucket of Pacific salt water on me. Needless to mention my crew wasted no time in meeting their responsibilities and took special pleasure and enjoyment in this tradition. Helen took a picture to capture the moment.

A Rally dinghy met us at the Balboa Yacht Club, guiding us to an assigned mooring. Safely secured, we enjoyed sundowners and recalled the highlights of the day realizing we had fulfilled every sailor's dream of transiting the Panama Canal. Panama City is a beautiful modern city with its Colonial District, Original City ruins, and Modern Panama City complete with skyscrapers, malls, and wonderful restaurants.

Many Panamanians express their sadness about the US leaving the Panama Zone and the many benefits including jobs that disappeared after the turnover.

Many Panamanians speak some English due to the US being in Panama for ninety years.

At the Balboa Yacht Club when trying to take on diesel our starboard fuel tank began leaking. Knowing we had less than two days to complete repairs I contacted a local welder. After several attempts to repair the tank he informed me he burned three additional holes. He gave us the name of a German man who he recommended could repair our tank. He was a colorful fellow, with the longest beard I'd ever seen on a man. Cutting a large hole in the top of our tank he began welding, then filling the tank with water to test the tank.

The man was clearly an artist and determined to repair my tank no matter how long it took. At ten o'clock his wife and daughter came aboard, the daughter falling asleep topsides. The family reminded me of our own hippy days in the 70's. At three a.m. in the morning the repairs were completed and we were ready to leave on schedule with the other Rally yachts.

The trip from Panama to the Galapagos Islands is 933 nautical miles and took seven days to reach this special archipelago. On the evening of the third night there was no moon or light excepting our mast light. Boobie birds with white underbellies were circling our boat like ghosts aimlessly darting overhead. The affect of seeing only their white underbellies created a surreal affect that entertained us for hours.

On the fifth day we crossed the equator at 0648. I was King Neptune and Helen Queen Neptune who held court on our subjects, the crew. In recognition of this first time event the Master of the SV *Tahlequah* did here insure proper and adequate libations, beer before breakfast. This event was duly recorded in the Ships Log. Furthermore, it was recognized that each crewmember did behave in such a foolish manner as to satisfy all traditional requirements by said authority, Helen and me. Over the Single Side Band Radio we enjoyed listening to other yachts entertaining and colorful originality, some of which can't be repeated.

At daybreak on March 6th we sighted the Galapagos Islands. These islands are considered part of Ecuador and the Ecuadorian Navy is responsible for their protection and administrative function. Their natural volcanic beauty is breathtaking; we knew we had come to a very special place. Visiting yachts in the Galapagos are restricted as to where they can venture and must be accompanied by special guides to explore the out islands.

We anchored in Puerto Ayors, Santa Cruz Island. Before arriving we signed up for a four-day special tour aboard the boat, Golondrina. The Golondrina was a vintage but wonderful boat that accommodated eight guests with private cabins. We recommend to our friends this is a once in a lifetime experience that everyone can appreciate and enjoy.

Imagine being in a place where the lion and the lamb live together harmoniously and fear nothing, this is Galapagos. Imagine walking along a beach with thousands of sea lions resting, sleeping, playing, and cavorting in the sand without fear of humans walking through their midst. We found the same true of all life in Galapagos including birds who landed at our feet, iguanas who sat staring up at us, penguins who took great interest in us, mating blue footed boobie birds who did their mating dance ignoring our presence as just another form of life among them. Nothing we encountered in the Galapagos Islands had fear of humans; they accepted our presence with interest and without alarm.

The first day of the tour we motored to Santa Fe Island, anchoring in a small but very beautiful bay with turquoise waters. Large schools of fish swam along the bottom as did rays and sea turtles. Sea life was abundant here and took interest in our presence. We were provided snorkel gear and taken to a place where there were many sea lions. The moment we entered the water the sea lions wanted to play with us. They nibbled on our fins, darting to within inches of us, challenging us to reach out and touch them.

Each day we visited another island (often eight hours apart) including Espanola Island, Floreana Island, North Seymour Island and Baltra. Each island was noted for different marine, bird, and mammals. The endemic species of these islands have been threatened by the introduction of wild goats, pigs, horses, rats, and cats through the last century. The National Park Authority has undertaken a program to eradicate these species in an effort to restore the balance of nature in the Galapagos Islands.

Snorkeling off Turtle Rock on Espanola Island I saw two six foot long white tipped sharks below me. I lay motionless, watching them circle fifteen feet below. Although I took interest in them they seemed to pay me little attention and continued their morning excursion on the reef. Once back aboard *Goldandrina*, I was informed by our guide Raphael these sharks although large were not known for attacking man. This didn't stop me from counting toes and fingers.

We sailed to Post Office Bay on Espanola Island, the oldest still operating post office in the world today. Here in 1792 whalers and pirates left a barrel ashore for mail to be taken by other seamen to the desired recipients. Sometimes taking many months to get mail to family and loved ones back home in countries all over the world. This practice continues to the present day (encouraged by the National Park Service) and is presently supported by both sailors and tourists.

Everyone leaving mail attempts to find mail they can return home with to be delivered as in days of old. In addition to the barrel sailors left oars, engraved wood, carvings, clothes, and articles of all kinds to commemorate their arrival here.

The natural beauty of the islands is breathtaking and the wildlife is so prolific it's impossible not to be impressed by the experience of being here. Swimming with sea lions, giant rays, sharks, and every type of fish imaginable each day adds to this remarkable experience. The nesting and breeding grounds of the blue footed Boobie Birds and Albatross is another remarkable experience that allowed us the brief connection between man and nature.

It occurred to us that in other lands the fear between animals and man is a self taught and learned one, animals have learned to fear man for good reason. In Galapagos this fear doesn't exist and the connection between man and nature exists in a most pure and beautiful form.

On the final day of the tour, March 13th we returned to our own sailboat *Tahlequah* anchored in Puerto Ayora with other Rally Yachts. Being Helens Birthday we planned a special dinner and evening at the Red Mangrove Inn with our friends, Ivor, Bernice, and Crew (Safari) and Carolyn and Crew (Spicy Lady). After a wonderful dinner Helen and I retired to the penthouse that Ivor and Bernice had arranged for us.

Our friend, Carolyn of *SV Spicy Lady* invited us to visit a local club specializing in guitar music. We arrived at 2000, purchased drinks and sat at a table near the stage. The lights dimmed, the guitarist began playing Spanish classical music, and it sounded professional. The second half of the show he played and sang contemporary songs, also very good. This was a favorite local place where patrons kept cadence with the music and routinely cheered when the songs was finished. At the end of the evening, the guitarist sat at our table to talk with us. He enjoyed entertaining and loved the guitar. During this time we discovered he had a dental practice next door to the club. Carolyn and Helen made appointments to have their teeth cleaned the following day.

Knowing we must leave no later than March 18th for the South Pacific Marquises Islands we made last minute preparations aboard *Tahlequah*. This included food provisions and whatever minor repairs were required for our long twenty-four day journey. This passage would be our longest, 3033 nautical miles. Everything had to be planned in advance including navigation routes, spare parts, tools, and food required for a long trip.

Although we enjoy the company of twenty-one other Blue Water Rally Yachts on route to the Marquises we must remain self-sufficient. Until we leave, our time and effort will now be directed toward preparing to undertake our longest journey to date.

CHAPTER 12

We're surprised to learn that today more people climb Mt. Everest each year than sail around the world. This may explain why we see few sailboats in the South Pacific and only rarely encounter one at sea.

After taking on fuel, fresh vegetables, a hand of bananas and other last minute provisions we left Santa Cruz Island, Galapagos on the morning of March 18th for a 3033 mile long journey to the Marquises that would take twenty four days. This route is the longest of any required in our circumnavigation and has a reputation for being benign and pleasant.

Except for occasional squalls during the first ten days the trip lived up to it's reputation of being a pleasant one. Unfortunately during one of the early squalls we lost one spinnaker during the night and damaged our jib during a second squall a few days later. Sailing more conservatively we continued making good progress toward the Marquises. The benefits of radar and autopilot allowed us to have time for occasional DVD movies, reading, and Mexican dominoes.

Our main battle was with inconsistent wind, although carrying enough fuel to motor twelve hundred miles, we had to sail many days due to the distance regardless of conditions. For this reason we flew our second spinnaker frequently during the daytime. Each day we talked by Sat Phone with *SV Safari*. We looked forward to this contact each day; it frequently lifted our spirits throughout the long journey.

We reserved enough fuel to motor sail the final 250 nautical miles to our intended landfall, Nuku Hiva in the Marquises. Although the trip moved fairly quickly at first, each day's progress appeared to diminish. The evening before our arrival we celebrated with a festive dinner.

The last evening watch is always a special one, dotted with periods of reminiscing and appreciating the stars, which shine brighter, and the wind and seas even more beautiful. Towards morning the seas were especially flat, we saw Nuku Hiva rising up into the clouds from twenty-five miles away.

Blue Water Rally communicated to us through the morning net they wished us to change our landfall from port of entry to Daniels Bay, where we would anchor overnight with the fleet. Daniels Bay is where the TV series *Survivor* was filmed and clearly chosen because of both its natural beauty and remoteness.

Daniels Bay is breathtaking; we entered a large lagoon inside. Beautiful vertical green mountains cascaded from the clouds down into the lagoon. Several Rally Yachts blew their air horns welcoming us. Safari sent their launch to pick us up and take care of us for the remainder of the day. We swam, ate lunch, drank, had dinner, and drank more aboard Safari. It was a wonderful feeling to be among friends again and relax after our epic long journey of twenty-four days.

Daniels Bay was so named in honor of Daniel, a friend of all yachtsmen who offered travelers fresh water and hospitality. Anyone traveling in this area was aware of Daniel through the Cruising Guides, and remembered always this colorful character. It's unfortunate that the *Survivor* TV series relocated Daniel to another Bay that is inaccessible to sailboats.

The following day we returned to the official Port of Entry and checked in with the help of the Rally and were provided a briefing on the next leg of our journey to the Tuamotos and Tahiti. A French sail maker repaired our damaged sails; we could now turn our attention to enjoying this beautiful Polynesian Island. As we walked along the waterfront, children greeted us with a smile and hello as did adults. We noticed many here had tattoos, even the women. We later learned that the tattoo was born here in the Marquesas with the early Marquiseans, its history continues with today's generation.

At puberty, males began the practice of having their bodies tattooed. These tattoos detailed the history of major events in their lives; even today many males and females have traditional tattoos.

The Tiki god was actually a small child. The back of the stone statue was a couple making love. The turtle above them represented fertility and the front was a small child. Only in recent years have these people inhabited the shoreline as they were hunters, not fishermen.

We took a four-wheel tour of the island with a French woman guide. The island is lush with vegetation, very mountainous and the people handsome and friendly. We toured the entire island and went to the most remote parts. The Tiki Sites were of special interest to us. We focused on two particular sites, Kamueihi and Hikokua. These villages were inhabited through the early part of the twentieth century by the Marquisean people who followed their traditional ways. As the Marquises were extremely isolated and remain so even today things are slow to change here.

Because there were so many early inhabitants consisting of many tribes they became fierce warriors. Cannibalism was a ritual practiced by the village Priests and even today many bones are discovered buried beneath the roots of the sacred Banyan Trees.

After spending a wonderful week in Nuka Hiva we went to the neighboring island of Ua Pou and Hakaketau Bay. Here we remained two days and were greeted by the friendliest people we have met to date. Because few sailboats come to these islands they took great interest in us. Everyone, young children and adults came to us and said *hello!*

Although the village was small we passed an open structure with people inside listening to music. Standing in the doorway people welcomed us inside to sit and listen to their music. We were delighted to sit listening to the music and watching the many children playing. People offered us food; a teenager without prompting came to us and welcomed us to their island in English. Later we returned to our boat to retire for the evening, grateful we had stopped here to visit.

The national pass time in these islands is canoe racing, we saw them everywhere. Young people practice rowing day and night hoping to improve their skills. The following morning was Sunday and we decided to attend the local Catholic Church regardless of the fact we couldn't speak Polynesian or French. Seems the entire village turned out to attend Mass.

Women were dressed in beautiful colorful pareos with flower wreaths on their heads. Young girls wore flowers behind their ears as did some men. Some distance from the Church we heard Polynesian songs being sung by the congregation. Polynesian songs are very melodic, beautiful, and extremely moving.

Although we had attended Catholic Mass before, this was unlike anything we had seen. Everything in the Church was hand carved including the pulpit, which was a ships bow. The tabernacle was a beautiful rosewood tree. It was an open church without stained glass windows. Only the natural beauty of the mountains outside could be seen from inside the church.

We were amazed at the number of teenagers, young children, and families in the church. Women played a very active role in the mass. There was a continuance of Polynesian Singing throughout the mass that only ended at the final moments of the celebration. Teenagers sitting next to us were singing aloud tapping their feet and hands in a wonderful rhythm that kept pace with the words of their Polynesian Songs.

I noticed during the entire mass small babies were being passed about from one person to another. The Polynesian woman sitting in front of us turned around and handed her baby to Margot from *SV Dr. Byrd*. After she held and cuddled the baby for a while Margot then passed the baby back to its rightful owner. I left that church feeling these were the friendliest and most welcoming people we've met anywhere. Although we wished to stay longer on this island we were limited to two days as we now had a four-day trip to the Tuamotus.

Because the Tuamotus are among the most dangerous atolls in the world for sailing vessels we had to carefully prepare our navigation route, double-checking again and again. We planned our entry to the Atoll of Fakarava to be slack tide during daylight hours. Upon arrival we discovered the current was running 5 knot's against us and decided to wait until the situation improved. In the company of another boat *SV Gee Wiz*, we delayed some time; we finally decided the time was now or never. Going through the reef we made only one knot over ground, once inside the water was smooth, protected and the Atoll beautiful and serene.

We learned later that because of irregular weather conditions and an atmospheric low, the atolls were dangerous to enter at this time Although we planned to visit several other atolls we quickly decided we wouldn't risk another entry elsewhere, but instead remained here until leaving for Tahiti.

Few people on this island speak English as their primary language is Polynesian and secondary language French. There is a small population in Fakarava; however the French built a school, medical clinic, small post office and even a concrete dock. A supply ship comes once a week bringing much needed supplies, fuel, and mail. Everyone gathers at the dock to claim their deliveries and watch the activity centered on the waterfront. Everyone smiles and is friendly and helpful. Regardless of the language difficulty, Helen found a method of communicating with the local women.

A lucrative French man established a black pearl factory here, wholesaling black pearls to Tahiti. Although it's a small operation with only four employees it's a fascinating process to observe. We're told it takes one and a half years for a black pearl to mature and less than two percent are considered perfect black pearls. Artificially inseminating oysters with seeds is comparable to a medical type process executed with precision.

Sunday we again attended mass with the village. In honor of our attendance, they announced the sermon would be spoken in both Polynesian and repeated a second time in French. Although we don't speak French, we appreciated the honor and pretended to understand everything. Polynesian songs are so beautiful and were sung throughout the mass. This time I brought my digital pocket recorder to record their songs.

We had learned small cuts and abrasions can become serious in the South Pacific. Great care must be taken to avoid risk of infection, and early treatment must be provided. We had observed how minor cuts could become easily infected and how these resist the normal antibiotics we all carry aboard. Our good friend Paul aboard *Gee Wiz* during the course of a few days went from a simple break in the skin (knee) to having to have a doctor cut out the infection, which was eating its way to the bone. Fortunately, a French medical doctor who sailed here 10 years ago stayed in Fakarava to provide much-needed medical attention.

CHAPTER 13

For several months, we discussed the need for long-term reliable crew and considered the complicated issues involved with having a third person aboard. Even a professional crewmember becomes an extension of our family, it must be someone who complements us, knows the basics, and is dependable. Observing several boats in the Rally having problems with crew, we didn't want to repeat their failures. Although Helen and I previously did long legs without crew, night watches alone are grueling with three hours on and three hours off.

Insurance companies charge additional hefty premiums with two aboard on trips in excess of 250 nautical miles from shore. Considering the possibilities, Michael, our grandson, came to mind. He was 22 years old, strong, good-natured, and always willing to learn. Imagining Michael aboard, we felt he would benefit from the experience, and in the end, we would benefit.

We contacted Michael, making him an offer on condition he would stay a minimum of six months. He could then make a final decision whether to continue or return home. To our delight, Michael accepted the offer; we purchased his airline tickets and awaited his arrival.

French Polynesia is a vast expanse in the South Pacific consisting of hundreds of islands; many are sparsely populated. The islands from Tahiti to Manuae, including Bora Bora, are referred to as the Society Islands. Four days after leaving the remote atoll of Fakarava, we arrived in Tahiti, a bustling metropolis that has become the focal point for cruise ships, tourists, and the capital of French Polynesia. Most goods shipped to French Polynesia enter Tahiti, and then flow to the hundreds of surrounding islands.

On the morning of May 1, we were cleared by the Port Captain to enter the port of Papatee (the main city in Tahiti), and instructed to tie alongside the new water front quay reserved for the Rally. A last-minute squall prevented our entry, so we remained at sea until it passed. We were greeted by Tony of the Rally Advance Team, who guided us through the necessary Immigration and Customs requirements.

Later that evening, we met our grandson Michael Fonseca, our new crewmember aboard *Tahlequah*. Apparently Michael hadn't received the information we provided earlier concerning the name of our yacht and its location. Customs refused to allow him to enter, and in frustration, finally escorted him to us in the main terminal demanding an explanation. Mentioning the Blue Water Rally carried a lot of weight. Customs released him into our custody. Once again we were a crew of three, adding to the camaraderie and fun aboard *Tahlequah*.

Unlike many of the smaller islands we visited, Tahiti is no longer the Hollywood version of the South Pacific. Although island geography remains beautiful, the city of Papatee is overpopulated with people, cars, restaurants, businesses, and tourists waiting to board cruise ships. The Tahiti Tourist Board provided free tours of the island and an evening of entertainment with real Tahitian dancing and music. Helen enjoyed the Polynesian men dancing as much as Michael and I enjoyed the women. A few days later, the Tahiti Yacht Club provided a luncheon for the Rally, which we all enjoyed.

Journalists and newspaper photographers began appearing at Blue Water Rally events, interviewing and photographing us. We saw a front-page newspaper article titled "Circumnavigators," showing pictures of Rally participants and events. Our new status was diminished as we went back to repairs and washing the boat.

On May 9, we left for Moorea, another beautiful island, and we anchored in Cook's Bay where the Rally planned another rendezvous. A small hotel complemented the natural beauty of the bay. The hotel provided us an evening of crab racing, and a few nights later, the hotel provided spectacular Tahitian dancing and music.

Like most French Polynesian islands, Moorea is surrounded by a reef. Helen, Michael and I went on a snorkeling trip, seeing many sharks and large stingrays. Stingrays came to us wanting to be petted. The more we stroked their heads, the more they made attempts to get attention. They are the gentlest of all creatures we've met to date. One pregnant ray allowed us to rub her belly gently, clearly enjoying the attention. Polynesians believe rays represent the spirit of their ancestors and won't catch or harm them.

One evening, we participated in a dinghy drift. Each crew brought food to share as we drifted through the anchorage munching on goodies. We were treated to a magnificent sunset; eventually drifting towards the open sea, we abandoned our partying. Each week, we continued attending a local church to hear more Polynesian singing. Each time the congregation repeats the service twice, once in Polynesian and once in French in our honor.

We were attempting to develop convincing expressions portraying our understanding of French, which went right over our heads. Our last experience, the minister offered to repeat the service three times, once in Polynesian, once in French, and finally again in English, but we declined and said French would be just fine for us.

We left Moorea for Huahine (pronounced Wa- He- Ne) and arrived at daybreak following an overnight sail. We anchored in a remote anchorage inside the atoll between the main island and a narrow strip of land. The natural beauty is indescribable, a white sandy shallow bottom, beautiful light blue crystal clear water with a backdrop of beautiful sandy beaches, lush green mountains and palm trees. Finally, we discovered the Hollywood version of the South Pacific.

We dinghied to a beach and were met by local people who invited us to use their beach, grass huts, and barbeque pit. They told us few boats came here and welcomed us with open arms and hospitality. We promised to return that evening for a beach barbeque.

That evening crews from the four boats returned to the beach as promised. There were two local Polynesian women swimming topless with their children. Trish (*SV Mizu Baby*) and Helen swam topless also. The local family living there caught many fish speared earlier in the day for us, and they also provided coconuts for drinking, and so our fish fry began. We knew offering money for the fish would be considered an insult; instead, we coaxed them to join us and share the food. At first they seemed shy; the women on the Rally personally invited the Polynesian women. This worked; they joined us with their families.

Running low on fish, a young man took his net and returned a few minutes later with more fish. No scaling and filleting here; they throw the entire fish into the fire, and you have a delicious fish dinner. We returned the following night, and once again they provided the fish, and we repeated the success of our first encounter with the local family. One young Polynesian couple saw me heading directly for a reef on our return to the boat in the dark. They sprang into their boat and raced toward our dinghy forcing us away from the reef.

We continued to notice the beautiful Polynesian tattoos mostly on men and occasionally on women. Although the men's tattoos consisted of ancient Polynesian symbols depicting their lives as fishermen and farmers, the women had small but attractive ones on their olive skin. These tattoos were distinctive and cannot be confused with any others. The symbols for a fish, whale, ray, shark, sun, moon, and life are almost petroglyphic and have deep meaning representing the heritage of a proud people. Seeing my tattoo, men sometimes asked where I got mine; I'd proudly respond *Nuka Hiva*.

I tried my hand at scrimshaw, but it's beyond my capabilities, so instead I've returned to doing pen and ink drawings in my log as a record of our explorations.

Having a tattoo in Nuka Hiva is a sailor's rite of passage, and many choose a traditional design. The first decision to be made is the traditional method using a shark's tooth and wooden mallet, or a needle. Deciding upon the needle, I sought out a recommended young man known for his tattoo skills living back in the jungle.

Arriving there, he instructed me to remove my shirt and lie down in a hammock. No patterns or drawings were used. He sketched an iguana in a circle on my arm, the traditional sign of good luck.

During this process, I asked where he learned this trade.

"My mother," he replied.

His young mother appeared; the entire left side of her body was tattooed. When I asked who tattooed her, she replied "I did; I'm right handed." This explained why only the left side of her body was tattooed.

When I asked her where she learned these skills, she replied, "My grandmother taught me."

Once you begin a tattoo it's too late to stop, no matter how painful. I purchased beautiful hand-carved necklaces made from dried coconut shells from his mother.

Before leaving the lagoon, we considered as a group what to do in return for their friendship and hospitality. Knowing we couldn't offer money, we considered fishing supplies. Some boats had fishing hooks, and we had our new Panama Warps, and every boat had some surplus item they could donate. One of our skippers put everything into his dinghy and brought it to the island. Needless to mention, our gifts were accepted with much appreciation.

As sailors, this interaction between us and local families were one of the highlights of our trip to date, an experience we will carry with us always. It was nice to be in a place where stealing and crime are not a consideration when traveling. We no longer had to raise our dinghy on the davits at night or put in safety bars on our hatches. Stealing simply was not a way of life here.

We were now en route to an atoll with two islands inside, Raiatea and Taahah. These islands can be described as the sailing centers of Tahiti's Society Islands. We went to the municipal marina in Raitea to re-provision and do laundry before going to the neighboring island of Taahah. Here we awaited the arrival of Ivor and Bernice. It was good to see them again, and we celebrated our reunion at a local restaurant on the evening of their arrival. As there were reefs and coral heads scattered throughout the atoll, we went cautiously to our next anchorage. Here we went spear fishing but came home empty handed. We did see beautiful fish on the reef.

It was time to head to our final destination in French Polynesia, Bora Bora. During the trip, our autopilot failed, requiring me to fly back to Tahiti to have a part repaired. This completed, we were able to enjoy the beautiful island of Bora Bora, which could well be the most beautiful and exotic island in the world. It's an island protected entirely by a reef and has the clearest waters we've seen anywhere. The colors range from a beautiful light blue to a dark pacific blue with white sandy beaches.

Unfortunately, Bora Bora has surrendered its shoreline to mega hotels that built exotic-looking chalets on the water numbering in the thousands. Bora

Bora is the most expensive place in the world. A room here in a local beachfront hotel costs $1,000 to $2,000 per night. Japanese come here to honeymoon and vacation. We wonder how local Polynesians have benefited (if at all) from big business development in Bora Bora.

While in Bora Bora (most western of the Society Islands), we met a successful businessman, who owned several pearl farms throughout the South Pacific. After doing us several favors, he asked in return to join us on the next leg of our journey. Not knowing him well and having past experience with inexperienced crew, we turned down his request saying, *with Michael aboard, there was no room.*

When returning to our boat the following day, we found our deck garnished with baskets of fruit, a gift from our new friend. A few days later, he returned to give us a tour of the island aboard his powerboat. At the end of the day, his girlfriend asked Helen if we would be willing to take a package; without waiting for an answer, he handed Helen a heavy package requiring two hands to hold it. Inquiring what was inside the package, he informed us it was black pearls, and he would pick them up at our next port of call. Returning the bag of black pearls, we left Bora Bora without seeing our friend again.

In leaving Bora Bora, we completed our tour of French Polynesia. In retrospect, we wish we had more time to explore remote islands and anchorages before leaving this wonderful area of the South Pacific.

On June 6, we left Bora Bora for an eight-day trip to Niue, a remote island en route to Tonga.

The trip from Bora Bora to Tonga is 10 days, the next rallying location for the Rally Yachts. Some yachts decided to go to Raotonga, some Aitaki, some Nuie, and others directly to Tonga.

Based upon weather models, we made a last-minute decision to go directly to Niue, a New Zealand protectorate located 1060 miles west of Bora Bora and only two days from Tonga. We raced to Nuie hoping to arrive there before the area was threatened by a low moving in that direction. On the fifth day, we began to see an increasing number of squalls, many we avoided thanks to radar, and others we battened down to make way.

On the sixth day, a military plane buzzed us at 30 feet above the water. He radioed:

Plane: "*Tahlequah, Tahlequah*. This is the New Zealand Royal Air Force; do you copy?"
Reply: "NZ. *Royal Air Force*, we copy. What can we do for you?"
Plane: "*Tahlequah, Tahlequah*. Just checking, and would like to ask a few questions."

Reply: "N.Z. Royal Air Force, no problem; happy to cooperate."
Plane: "*Tahlequah, Tahlequah.* You're out of San Francisco. Where are you headed?"
Reply: "N.Z. Royal Air Force, coming from Bora Bora and headed to Nuie."
Plane: "*Tahlequah, Tahlequah.* I love French Polynesia. What is your Nuie ETA?"
Reply: "N.Z Royal Air Force, we love French Polynesia, too. Niue on 6–14 at 1400."
Plane: "*Tahlequah, Tahlequah.* How many on board and captain's name for our records?"
Reply: "N.Z. Royal Air Force, three persons and captain is Echo, Delta, Mike, United, Echo, Sierra, Charlie, Hotel." (Ed Muesch)
Plane: "*Tahlequah Tahlequah.* Thanks, mates, and you blokes have a safe trip now. If you need anything, give us a call on 2182 kHz, and we'll be there."
Reply: [Thinking: Wondering if this includes a beer run? Keeping thoughts to self] "NZ Royal Air Force, thanks for the offer. It's appreciated."
Plane: "*Tahlequah, Tahlequah.* You blokes take care now."
Reply: "N.Z Royal Air Force, CLEAR!"

Two days later, we found ourselves in increasing high seas and winds. Although winds topped 35 knots, they would increase to 45 knots within 24 hours. We expected to arrive in Nuie as the frontal system passed. The remainder of the fleet was behind us and would encounter the full force of the front.

The following morning, we saw Nuie for the first time as the sun rose over the cliffs. The cliffs made of limestone were 150 feet high and seemed to encircle the entire island. On arrival, we radioed Radio Nuie to announce our arrival and request Immigration and Customs to clear us. After boarding *Tahlequah* and performing a cursory search, we were officially welcomed to Nuie. They got few outside visitors since a disastrous cyclone six months earlier. The following day, four other Rally yachts joined us.

Going ashore, we saw the full extent of the damage to the island. Although the cliffs are high, the tidal wave caused by the cyclone was higher and did tremendous damage to the island. The hospital, hotel, homes, museum, and stores were swept away or heavily damaged. Only 1,100 persons today live on Nuie, and although New Zealand continues to provide foreign aid they no longer officially qualify. Nuie has its own language, and to us sounds Polynesian.

Walking through town, everyone waves or stops to talk, asking which boat we're from. They take great interest in our stories and life upon the sea. The local yacht club now operates from a small storefront. They informed us we would be guests at a barbeque in our honor. A few days later, we attended the barbeque, meeting many of the local townspeople who could not have been more friendly and gracious.

We became honorary members of the Nuie Yacht Club and promised to spread word of Nuie among the cruising community. Locals offered rides to transport us wherever we needed to go. A local dentist rebuilding his clinic offered us a ride helping to transport jerry cans to the only gas station on the island. Without asking, the attendant put our jerry cans on his truck and transported them back to the wharf refusing to accept money.

The following day, we rented a car and drove around the island. Town after town was entirely abandoned except for one or two dwellings. Strangely, we saw two newly constructed churches, one Pentecostal and one-Seventh Day Adventist Church, built amidst the ruins of an abandoned town. Although there was no one to attend these churches, they were built with donated funds from other countries.

We stopped at overlooks with signs explaining the history of the site below. We were saddened to discover stone walkways disappeared at the edge of the cliff as the cyclone swept them into the sea. There were few limestone caves and pools left, and they were also completely destroyed. One sign described a cave with beautiful bathing pools used by the Kings of the past for hundreds of years, but this, too, was lost.

We visited the Washaway Café, seemingly one of the few structures to survive the cyclone at water's edge. We met the proprietor earlier at the Yacht Club and promised to visit his Café. It reminded me of the stone dwellings I was familiar with growing up in Arizona. The entire café was built from large coral and limestone rock with inlaid beams open to the sea.

He cooked us the best steaks we've had in some time and sat with us the entire evening to discuss Nuie culture and his part in helping to save the island. As he looked like Cheech (from Cheech and Chong), I tended to take every word as truth and soaked up his ideas the entire evening.

We were invited to our first rugby game at the old Rugby Club Saturday afternoon. People offered us rides, taking us to the clubhouse, where the game was already in progress. At the end of the game women sold island food.

After farewells, we were once again on our way, this time to Tonga. The first night of our voyage, we saw the beautiful Southern Cross in the heavens—a sign we would enjoy a safe four-day voyage to Tonga.

CHAPTER 14

En route to Tonga, we were within VHF radio range of other Rally Yachts. Three days from Tonga, we heard a radio transmission alerting all mariners that the King's Private Ceremonial Canoe had broken free of its mooring and was now adrift at sea. The Tongan Royal Family was offering a reward for its sighting and return. Having never heard a Tongan accent, we had no reason to suspect fraud. We kept a watchful eye on the horizon. The following morning, another Rally Yacht radioed having a possible sighting of the royal canoe and provided coordinates to Radio Tonga who relayed this information to the authorities. During the day we heard several transmissions resulting in the location and return of the craft to the Royal Family.

The following morning, the authorities requested the ETA and Port of Call of the yacht having located the Royal Canoe. This established, Radio Tonga stated the King would personally meet the yacht and a gift of land provided. One Rally Yacht radioed to congratulate our lucky fellow companion. Within an hour, the clever couple that orchestrated this folly admitted to the hoax. Although every detail including accents seemed authentic, a number of us were fearful of being duped. During long passages, it's important to rely upon someone for entertainment. This couple enjoyed a reputation for fun and originality, and they could be relied upon to always come up with something to keep us all going.

At daybreak we sailed the northern island group of Tonga (The Friendly Islands), Vav'au entering the port of Neiafu in Refuge Harbor. Tonga consists of three island groups and is ruled by the King of Tonga, also head of the Methodist Church. The Rally requested we proceed directly to the customs and immigration dock to await clearance before taking a mooring inside the harbor. After tying alongside the wharf, customs, immigration, and health officials boarded *Tahlequah*. They were dressed in traditional clothing consisting of a long dark sulu (dress) and straw mat tied about their waist. Many workers and children wore similar dress, and it was apparent this was a far different place and people from anything we previously encountered.

The Rally was treated to a Tongan feast and dance by local students at the Paradise Hotel above our anchorage. Tongan dancing is different from French Polynesian dancing, as the movements are slow and show hand expression. Their bodies are covered with coconut oil and glisten in the light. We took many pictures and recorded their beautiful Tongan songs.

It was here we experienced our first Kava drink when a group of local men offered us theirs. To benefit from the effects of the Kava root, large quantities must be consumed. On this occasion, we only tasted Kava, a repulsive drink unless you're used to it. It appeared to us to be a male social ritual going until the early hours of the morning. It can produce a drug-like state, and in smaller quantities produces only a relaxed feeling. Strangely, our women who partook of Kava claimed they experienced no side effects. Men described their lips going numb after the first cup. In the past it was the woman's function to chew the kava root, spitting the contents into a bowl that the men would later drink. Needless to mention, our pleasure to learn this custom was abandoned in favor of more modern methods.

Sunday, we attended a local Methodist church; it was crowded with several hundred parishioners dressed in sulus with straw mats about their waist. The influence of the church is apparent; women swim fully clothed and scanty attire is frowned upon if not prohibited. Nowhere else was the influence of Christian missionaries felt so strongly.

We sailed to other nearby islands where the Rally organized more Tongan feasts and dancing. The food was local, consisting of raw and cooked fish and much finger food that is hard to describe. The people were shy, but always friendly and hospitable. One evening a village displayed beautiful shell jewelry, hand-made woven baskets, woodcarvings, and hand-painted Tappa Cloth with no pressure to purchase anything. Tonga is famous for Tappa Cloth, which is made from the mulberry tree, then beaten by hand until flat and dried in the sun, and then hand painted with traditional Tongan designs. Although we enjoyed Tonga, it proved too conservative for our liking, and at times we felt the people were unable to freely express themselves.

We left the Kingdom of Tonga for Fiji, a three-and-a-half day trip. The first night the autopilot broke, leaving Michael and me to hand steer. Because of the difficulty, we reduced watches to two hours instead of three. We sailed for the port of Savu Savu on the island of Vanua Levu. We anchored in front of the Royal Suva Yacht Club where the Rally arranged for use of their facilities. The hospitality of the Yacht Club was beyond our expectations.

One evening the town's mayor in traditional dress greeted us at a special cocktail party given in our honor. Again, young adults and children performed traditional Fijian dances to entertain us. Each day, the Yacht Club organized events, which included tours, dinghy sailboat races, and music. The final day of our stay in Savu Savu, they constructed an open hut of palm tree branches and a pit dug for cooking food in the traditional manner. A large crowd in traditional dress gathered on the grounds of the Yacht Club.

Straw mats were scattered about, and the men began to brew Kava for our Kava ceremony, a custom we first encountered in Tonga. The custom requires a large wooden bowl with Kava Root wrapped in cheesecloth soaking in cold water. Traditionally, this is a custom for men; however, our women partook as well. We were informed the man sitting in the palm hut was the Tongan Chief of Chiefs. He was here to welcome us officially to Fiji. Although prohibited from turning our backs to him, each of us stood on line to be personally greeted.

He asked Helen and me where we came from, the name of our boat, and shook our hands, appearing to enjoy the experience, as I did. He was surrounded by a court of traditional-looking warriors in palm leaves and grass. A local New Zealand sailor living in Fiji explained the significance of each step of the ceremony, again emphasizing we not turn our back to the chief.

The chief's personal entourage of men and woman dancers performed for us and were delightful. The women and men were clearly middle-aged, unlike the body-beautiful dancers we came to expect in Tahiti. This made them no less impressive; they were athletic, quick, and were very proud. A number of us purchased Sulus (long dark skirts) for this ceremony. We were told as foreigners our sulu dress was seen as a sign of respect.

At the Kava Ceremony, each of us was presented with a coconut shell filled with a liquid looking and tasting like dishwater. The man who passed us the kava clapped once. Upon drinking the kava, we were required to clap three times in response. The first drink of Kava left my pallet numb, a sign everything was going well. Drinking more kava left me with a strange, indescribable feeling. Today, when visiting surrounding islands, it's first required you visit the chief, presenting him with a gift of kava root. Only then can you be given permission to visit his island. Kava is a social custom many people enjoy in Fiji. Personally, we found it difficult to understand how they enjoyed this drink.

On July 4, we dressed *Tahlequah* in signal flags extending from the bow to the stern as did the other two American yachts with us. Later, the Yacht Club provided a true Fourth of July picnic, American style with hamburgers, and hot dogs with all the trimmings.

Planning a three-day trip, we sailed southwest to another Fiji island group, Viti Levu Fiji. On this occasion, we made the decision to leave the dinghy with outboard on the davits, something we rarely do. Without warning and after leaving the protection of the island, 35-knot winds and rough seas suddenly hit us.

The inflatable with outboard began dipping into the water each time we heeled. I realized if I didn't do something quickly, we risked losing our new outboard. Michael was bigger than me and feared the added weight would only contribute to the risk of losing the outboard. Although my hip was becoming a problem I managed with Michael's help to get over the stern rail, into the dinghy, and secured the motor lift halyard to the outboard. During this entire process, I was being banged about with waves washing over me. Once safely back in the cockpit, we raised the outboard securing it on deck.

Due to reefs, we anchored at night along the western coastline, taking two nights and three days to reach Musket Cove on the island of Naigainiand. Never before had we seen such breathtaking scenery. Having seen pictures of New Zealand, we could only imagine the landscape was similar. We saw high mountain ranges rising into the clouds above, and then cascading into flatlands several thousand feet below. The mountain ranges seemed to go on endlessly.

Arriving at Musket Cove Resort, we were again greeted by Rally Directors. We were encouraged to make use of all the Musket Cove facilities including, pool, sports equipment, dining rooms, and private beach. The owner, a New Zealander and sailor himself, greeted us at a cocktail party in our honor. Every day, there were planned activities, parties, and special entertainment, including a hobycat race we participated in.

Helen and several of the women went for massages at the health club; even I got a massage. Because you were presented with a Musket Cove credit card, money was never used to pay for items or services. Departing, you received a bill for total amount due. We maintained our own tab to track spending; in the end, everything was very reasonable.

One Rally couple, Margit and Benno of the *SV Dr. Bird*, (Swiss Registry) were married in Musket Cove in a traditional Fijian outdoor wedding ceremony. The groom and bride were dressed in similar colorful handmade tappa cloth with colorful hand-painted Fijian designs, traditional wedding clothes of the Fijian. That evening, we enjoyed dinner at the restaurant in honor of Benno and Margit.

This was a special day, not just for Benno and Margit, but for all of us in the Rally. Being away from family and friends for two years, Rally participants become extended family. Benno was always the even-natured guy who if he couldn't say something good, it wasn't said. He could be depended upon to make us laugh, especially when he was the radio NET Controller for the day.

Margit was in charge of midwives at a Swiss Hospital and although having medical problems she was the most loving, cheerful and active woman I'd ever met. We warmly thought of Margit as a mermaid, she would remove her clothes, jump into the water, swimming to each boat to greet us at the start of a new day. Margit always rushed up to me, kissed me on the lips, cheerfully saying "H … e..l..l..o, Eddie!" Although Benno and Margit spoke little English, and we spoke no German or French, it made little difference. They were among the warmest and sincerest people on the Rally.

We left Musket Cove briefly for annual bottom painting and routine maintenance, performed on the mainland. Because *Tahlequah* was on the hard, we stayed at the Anchorage Resort nearby the marina. Michael and I supervised the work while Helen enjoyed the pool and read her book. After four days, we were back at Musket Cove enjoying our good friends and resort facilities. For the first time since leaving home, we were on vacation. Michael took a diving certification course and became certified four days later. He was now qualified to clean our boat bottom!

Distracted, I'd forgotten to drink water for four days, believing I could substitute red wine. On the evening of the fourth day, I began showing signs of dehydration and went into shock sitting in the cockpit. I'd managed to ask Franco aboard Safari to get Helen prior to passing out. Claire of the *SV Condor* was holding my hand, saying *don't be afraid, you're going to be fine*. I'm thinking, why do I have to be sick at a time like this?

Dick and Leslie from the *SV Aragorn* gave me electrolytes, which helped rehydrate me. Fortunately, three women in the Rally were medically trained and began taking vitals and managed to get liquids into me. I was given a muscle relaxant and codeine for the pain. During this time, someone phoned a doctor on the mainland who confirmed both my diagnosis and treatment. He prescribed I lie down for the night and call him in the morning. By 2:00 am, the pain finally subsided and I was able to sleep. The following morning I began drinking water routinely.

On August 27, we left Musket Cove, Fiji, for a four-day trip to Vanuatu (New Hebrides). During the voyage, we encountered a severe lightning storm lasting the night, but escaped associated high winds. The yachts ahead and behind us were not so lucky and bore the full brunt of the weather front. Using radar, navigation software and our night vision scope, we entered the capital of Vanuatu, Port Villa, at midnight. Once inside, we anchored alongside the quarantine buoy until the following morning awaiting the arrival of Custom, Immigration, and Quarantine Officials.

We stopped at Vanuatu to reduce the length of the 1,750 nautical mile trip to Cairns, Australia. We heard from cruisers that Vanuatu was a wonderful country to visit and exemplified the South Pacific. Many of the islands in Vanuatu are extremely remote and isolated, and cannibalism continued in Vanuatu until the 1960s. The Capital, Port Villa is a typical expatriate community of New Zealanders and other nationalities. Arriving, we began making plans to visit the island of Ambrym, an isolated island with few outside visitors. We arranged for permission from a local Chief to visit the island. These arrangements were put together by Leslie from the *SV Aragorn*, another American boat.

Ambrym Island is referred to as the Black Island, not only for its two active volcanoes but its black magic. We went there to attend their Back to My Roots Festival, which is held annually between neighboring tribes on the island. The festival is not for tourists but locals to preserve their culture.

I met with the Vanuatu Air District Manager, and negotiated an unscheduled stop at an old World War Two airstrip; try that one with an American Carrier. Once we agreed on the diversion fee, we split the cost between nine of us. On August 4, we boarded a 17 passenger twin-engine aircraft and made the 40-minute flight to Ambrym Island. The passenger sitting next to me carried a long wooden spear, and to think I left my pocketknife on the boat. The plane landed on an overgrown runway consisting of broken tarmac and grass. Nothing was here, no people, no buildings except a small shed. As we disembarked with our backpacks, we were met by Isaiah, a local guide who would bring us to his village on North Ambrym for our stay of three days.

Near the airport were straw hut villages scattered through the jungle. After a 30-minute walk, we came to an old powerboat owned by the village, and we motored toward the north part of the island, where we stayed. Along the way, we saw many one-man wooden dugout canoes with outriggers on the beach ready for launch.

Two hours later, we arrived at the village, located above a rock cliff. Climbing down the cliff, men and children met us for the purpose of taking our backpacks and giving us a dish of sweet potato fries that we ravenously ate in 10 minutes. We then motored another 40 minutes through many reefs to a deserted beach. The sand was black and very fine, formed from volcanic rock and ash. We walked through the jungle for another 30 minutes passing more straw hut villages. Men, women, and especially the children waved to us.

We began to hear the sounds of totem drums and men's voices singing in harmony. Suddenly, two men appeared and informed us we would have to pay to go further. Knowing that all money goes to the chief to be used for his village, we were happy to pay. Moving on, the totem drums and singing grew louder and louder until we came to a clearing in the jungle. There were several large totems, hand-carved from trees, standing majestically.

To paraphrase Garrison Keeler's Lake Woebegone, Prairie Home Companion, Ambrym Island is where all the men wear penis sheaths (nambia), all the women are bare breasted, and all the children are beautiful and above average. The men wore only small straw woven belts with penis sheaths and a small fern through the back of their belts. Some men wore a long plume feather in their hair giving them a noble appearance. For the remainder of the day we were privileged to witness tribal ceremonial dances that ranged from Death of a Great Chief, to a Young Man's Coming of Age Ceremony. All 17 levels of chiefs were there, including their Great Chiefs. Some were very elderly.

The men were muscular and athletic, including the elderly. Men and women danced vigorously and chanted, beating their feet on the ground, moving in unison. Some of the men and women had red streaks of dye on their faces, giving them an intimidating and warrior-like appearance. There were many children from nearby villages sitting enthralled by what they saw. Many children carried long home-made knifes, not for defense but to cut and peel fruit in the jungle and open coconuts when thirsty. The children took great interest in us, as few white people came here.

Children took great delight when we acted silly, staring at us to see what we would do next. Michael showed the children his pierced tongue, entertaining them for hours. The language of Vanuatu is a pigeon English, and if spoken slowly can be understood even by us. A few examples of this are: *Pipe suckin wind* means snorkel, *glas* means diving mask, *me likem tu much* means I like you, and *Havem yu goodem* means enjoy yourself.

During the boat ride back to the village, I talked to our guide about a neighboring island, and his only response was *taboo, forbidden island, you no go there! end of subject,* and we discussed it no further. I talked to him about black magic, he later offered to take me to a sorsen (a sorcerer) in the jungle, something I thought well advised not to pursue. I discovered from Isaiah that each village has a chief; he makes all decisions for the village. Money is turned over to the chief, and he wisely distributes it as needed.

Arriving at our destination, we stopped below high cliffs, and young men from the village were there to assist us as we climbed to the top, not an easy task for me with a bad hip. At the top, we walked another 30 minutes past our guide's village. These were the simplest of straw huts, no furniture inside, no electricity, any rooms or dividers, just an open space with straw mats on coral to sleep on. Walking through the village, everyone waved and smiled. Begging is not a way of life here, offering a gift that person must give you something in return.

Each couple had their own private hut. Using dried coconut shells, village women heated water for us to bathe, a special treat. The first evening, the women cooked supper, consisting of chicken, sweet potato, and cabbage. The women, men, and children only ate when we finished. After dinner, our hosts laid a straw mat on the earth for us to watch the red glowing sky caused by the active volcano in the distance. Nine of us lay there mesmerized by the beauty of what we saw.

We brought gifts of rice, dried peas, beans, paper, pens, and toys, including a tambourine for the children. Helen presented the food to the women, and I gave the rest to Isaiah to be distributed to the village. We were asked what the women should do with the dried peas and beans; they had never seen them before. We explained how to prepare them, and hopefully they would be successful. Later that night, we heard village women singing to the children while a child kept beat with his new tambourine.

The second day of the festival proved to be better than the first. When we arrived, young boys (ages 6–13 yrs.) naked except for penis sheaths performed traditional games led by a Great Chief in his 70s. One game was Searching for Fire, and the boys laughed and enjoyed themselves. The youngest child was a boy too small for a penis sheath; instead, he was covered with a leaf. During the day hundreds of children sat enthralled and absorbed watching the day's events. There were demonstrations of magic, sand painting, and many ritual dances. The men dancing, stomping the earth in rhythm, and chanting to the sound of the totems was very moving. These were a very proud people. I took many digital pictures and short film clips.

At the end of the day, men danced holding three sacred ROM Dance Masks. These masks are sacred and can only be worn once, then destroyed, because it's believed they contain the spirit of the dancer. We were warned anyone seeing the masks being made or touching them would be heavily fined (to pay for the pigs that must be sacrificed), then whipped with a poisonous vine. Anyone coming close to the masks was warned away by their protectors, pretty serious-looking people.

Isaiah brought women from his village to prepare our lunch during the festival; we had fruit and coconut milk. As we sat on a straw mat to eat, the women stood by swishing the flies away using dried string-like vines.

At the end of the second day, Isaiah suggested we walk to a nearby village along the coast to meet the boat returning for us. An hour later, we reached a clearing with shade trees along the beach. Villagers were sitting on straw mats as children slowly came from the village and gathered near us. It was apparent few white people came here; we were of interest, especially to the children. We counted up to a hundred children, many sitting about staring at us.

Leslie, a woman from the *SV Aragorn*, stood and walked to the water's edge where she waded to a coral formation and sat. A young girl followed her and sat with her. Children from everywhere waded to join the young girl sitting with Leslie as she attempted to talk to the children; they only giggled in response. Finally Leslie asked a young girl if she knew what a doctor was, and she responded *no*. Leslie then asked if she knew what a medicine man was, and the young girl answered *yes*.

Leslie stood and walked back to the beach, beneath a shade tree. Suddenly there were many children sitting quietly around her saying nothing. As verbal communication wasn't working, Leslie attempted to teach the children how to play the game "Duck Duck Goose," and within minutes, the children were all playing, laughing, and having a wonderful time.

Our boat returned; it was time to leave. The entire village stood on the beach waving to us as we left, a very moving experience by anyone's standard.

When we returned to the village, a young man taught us to play the totem drum outside the village. We purchased wood masks, carvings, a walking stick, and small totems before leaving.

We've heard from other cruisers each island in Vanuatu is different and is a special experience. We felt we had experienced something special and were grateful for this wonderful opportunity. We will cherish this memory always.

Arriving back in Vanuatu we brought all our wooden carvings to an agency within the airport to be fumigated, a requirement to avoid confiscation in Australia. Because it was Friday, the officer in charge insisted it was too late and couldn't help. Unfortunately, we had to leave for Australia on Sunday and would have no other opportunity. Disappointed we all got into a cab to return to the marina. A thought occurred to me.

"Wait a minute," I said. "I'll be right back."

Returning inside to talk to the officer alone, I informed him we were willing to pay overtime if he would help us. He agreed if we would pay his cab fare to the marina. The total cost was $15, more than acceptable. We'll never know if our carvings were actually fumigated; however, we had a document allowing us to legally take them to Australia.

CHAPTER 15

On August 7, 2004, we obtained port clearance to leave Vanuatu for a 10-day trip to Cairns, Australia. Three days from Australia, we observed a pod of whales off our starboard side. There were young calves among the whales, one of God's most beautiful creatures; we didn't want to come between a mother and her calf.

The following morning, we received a fly over by the Australian Coastal Patrol; they verified our identity and issued specific instructions concerning quarantine and Customs. The final evening prior to our arrival was a beautiful clear evening, a soft warm breeze, and the Southern Cross overhead, the perfect setting to make love on the deck beneath the stars at change of watch.

August 17, we entered the well-marked Great Barrier Reef at night. We sailed north to Yorkies Knob Marina, where we were welcomed by Customs and Immigration. They confiscated our two eggs, one garlic, and a lone jar of mayonnaise. We surrendered these items willingly, and were delighted to keep other ship's stores we feared losing.

Overshadowing our trip since leaving Vanuatu was the lump Helen discovered in her left breast. The Rally received our e-mail from sea and arranged for a van to transport Helen directly to the Women's Breast Clinic in Cairns. Following an intensive physical, mammogram, and ultrasound, Helen was assured she had only cysts.

The Rally arranged for us to attend another mayor's party in our honor. We've met more mayors in foreign ports than Americans circumnavigating. Australia is the only country where everything from the largest animal to the smallest bug is intent upon killing us. Signs dot the beaches alerting boaters once landed to immediately move to the roadside to escape gators waiting for unsuspecting dawdlers.

After a few days, we relocated to the Marlin Marina in downtown Cairns. This did wonders for Michael's social life. The young women in the Rally always invited Michael when going out for the evening. Michael was a strapping fellow no man challenged; in his past life, he was a bouncer. To help Michael's image, I gave him boat cards saying, *Michael Fonseca, crew, bodyguard, and underwater repairs.* We enjoyed having Michael aboard as did other yachts. Michael befriended the skippers of other yachts, frequently spending evenings with them.

Within a short time of arriving in Cairns, Michael began experiencing severe headaches and nausea, especially during late afternoon hours. One day it became serious, and we rushed him to a medical clinic. After testing, we were informed Michael had malaria; fortunately, it wasn't the worst type. Taking him to the hospital at the clinic's advice, we received a secondary prognosis, it was malaria. The incubation period of 10–14 days placed him at Ambrym Island, Vanuatu, when he contracted the disease. Although we were taking malaria tablets, the doctor informed us there's no guarantee they work. The doctor prescribed medication and advised taking Michael back to the boat. There was nothing to be done for him there at the hospital.

As previously arranged, we would be touring Australia with visiting friends. Ivor on Safari volunteered to take care of Michael during our absence. Knowing Michael, I suspect Ivor threatened to tie him down. During the time we were away, we kept in touch with Ivor and Michael by phone. It took four days before Michael responded to the malaria medication. We're told the symptoms of malaria can return but are treatable with medication.

Our friends Debbie and Nolan arrived from the U.S. to spend 10 days with us. Debbie and Nolan have become routine guests aboard *Tahlequah* since meeting them at the Club Orient Naturist Resort in St. Martin. We toured Cairns, a wonderful modern city that attracts many young backpackers. There's a beautiful park with a lake-like swimming pool in the center. Young Australian sunbathers are everywhere, and topless is acceptable. There are many shops and restaurants for all tastes. We went on the *Ocean Spirit* (a local tourist boat) to swim and snorkel on the Great Barrier Reef. We took a submersible ride to fully appreciate the reef underwater, seeing beautiful coral formations and fish.

The following day we rode on the original Scenic Railway built in 1910 to Kuranda, high in the mountains. The mountains, waterfalls, and gorges were indescribable—scenery that can only be found in Australia. Nolan and I left the girls to purchase beer, finding a small bar on a back street. Inside, most patrons were Aborigines wearing bush hats, listening to Australian Outback music. Pool tables were scattered throughout the room; all were occupied. Two young white females sporting tattoos played pool with the men. Waiting for the woman behind the counter to get our beer, we stood there reading the list of names and pictures of patrons no longer welcome. Names like Jeremiah Wilson, Jesse Freelong, and Rebecca Free made the list. We purchased our beer and returned to meet the girls.

We stayed at a cabin within the Kuranda National Forest. It was comfortable and only a half-hour walk along the railway tracks from town. On one excursion to town, we passed a group of Aborigines, men sitting by themselves drinking beer and the women sitting on the other side of the tracks drinking soda. On our return, the Aborigines were gone, and we noticed the area appeared to have been cleaned and raked, and there was no sign of refuse anywhere. After dinner we entertained ourselves by sitting on the cabin deck watching wild turkey and wallabies.

In town we encountered a colorful fellow dressed like Professor Dumbeldore from Hogwarts in Harry Potter. He told us he enjoyed entertaining children and saw that as his calling in life.

Wow, I thought.

Although amused, we moved on, seeing the rest of the town. The town is a haven for tourists; we visited a butterfly aviary, learning all we ever wanted to know about butterflies. We returned to Cairns the following day, this time using the cable car instead of the scenic railway. It was a wonderful opportunity to take many aerial pictures of Cairns and the coastline.

We visited the Night Zoo in Cairns, seeing and petting many kangaroos, alligators snapping chickens from a bamboo pole, koala bears, snakes, and exotic birds. We were given Billy Tea (tea in a tin cup) and tucker (honey bread) while sitting around an open fire. Friendly kangaroos hopped about, willing to be petted in exchange for our food. Some of the kangaroos had small babies in their pouches. It was a wonderful opportunity to interact with these animals and an experience we won't soon forget.

After touring Cairns, we flew to Alice Springs, the outback of Central Australia. Alice Springs has a population of 20,000, many Aborigines, and retains a flavor of the true outback. The landscape reminds me of my childhood growing up in Arizona, arid and dusty. We stayed at the Desert Palm Backpackers Resort, having our own small cabin with kitchen. Evenings were spent on our patio, enjoying sundowners before going into town for dinner. One evening, we went to a saloon called Bo Jangles for beer and dinner. The sign said Free Peanuts, shells must be thrown on the floor, and they weren't kidding; we were up to our ankles in peanut shells. The local patrons were as bush-like and colorful as the wall hangings.

There was a motorcycle in a glass enclosure with a human skeleton; a large snake spiraled its way through the skull feeling quite at home. Nolan was taking pictures when a patron at the bar wearing a bush hat, told him not to take his picture. Not sure what that was about, maybe someone wanted for the big one. Alligator hides, kangaroo pelts, and other indescribable items decorated the walls and ceilings of this popular town hangout. Standing at the urinal, I was forced to make eye contact with a buffalo conveniently positioned to keep me on my toes.

We visited the Royal Flying Doctor Service founded by the Rev. Dr. John Flynn over 75 years ago. In the museum, we saw films and displays showing how medical persons fly long distances throughout the outback to bring medical help to those needing it. Today Flying Doctors are always there, 24 hours a day, providing a health service to people in more than 80 percent of Australia. There were displays of historic dental and medical equipment and communication equipment used by the Flying Doctors before the telephone was widely used.

Models of original and modern aircraft hung from ceilings. Today, the Royal Flying Doctor Service provides comprehensive health care to people who live, work, and travel in outback and rural Australia. Visualizing the great vastness of the Australian outback, one can appreciate what these men and women contributed.

Next, we visited the Radio School of the Air founded in 1955, where teachers provide an education to children using a radio and communicating with them in their farms and homes. A hundred and forty children living in isolated settlements covering over one million square kilometers of Central Australia are students. All families are given transceivers and provided their own frequencies. The only friends students have are those they talk to during lessons. It was exciting to watch and listen to teachers and students interact through radio in the same way any normal school does.

One change will soon occur, bringing students into the modern age. Radios will be replaced with computers so teachers and students can talk and see one another. Although students can only progress to the seventh grade, many go to boarding school for continued education. The Radio School of the Air each year sponsors its students to come to Alice Springs to meet friends and experience city life.

We visited the Aboriginal Culture Center, founded by a man half aboriginal; he's considered one of the Lost Children. He explained he was taken away from his mother as a child by the Australian government because his father was white. Placed in a white home, he grew up receiving the benefits of a white child. Although having difficulty being accepted back into the Aboriginal community, he talked about re-entering the Aboriginal culture and way of life. The practice of creating *Lost Children* continues to be a controversial subject in Australia.

He took us to the old Telegraph Station, where Alice Springs was born, explaining to us how Aborigines lived in the area. In the bush we saw many kangaroos that took great interest in us from a distance. Both Nolan and I bought Didgeridoos, the Aboriginal instrument having a unique sound. Although we took lessons and are considered graduates of Didgeridoo U, only Debbie showed promise.

It's difficult to visit Alice Springs and not have impressions and form opinions of the Aborigines. We saw large groups of Aboriginal women and men gathered on lawns of public buildings and parks, sitting in circles. We appeared to be transparent to them; they paid no notice to us, keeping to themselves.

It's important to first understand they are one of the oldest civilizations in the world today. Their religious principles are based upon Dream Time. The ritual definition meaning the time before times, from this much of their beliefs and practices are based. Dream Times establish an important relationship between the physical, the human, and the sacred, and are the basis of their spiritual understanding of the universe and many of the laws governing their social behaviors.

I asked Helen why they appeared to ignore us. Helen responded it must be the result of how they have been treated. Not long after, I was reading the log of Captain Cook, one of the early European explorers coming to Australia. He commented that when coming ashore, the Aboriginal villagers ignored him. Only when he went into their village did they take notice, attempting to frighten him away.

I saw many shops in Alice Springs claiming to help Aboriginal Peoples. These included the Aboriginal Divorce Assistance Program, Aboriginal Legal Defense Network, and Aboriginal Psychiatric Service. I couldn't help but wonder what these western solutions were doing for a race successfully existing over 20,000 years without the white man's influence. I wondered how potentially destructive these well-meaning people were to the Aboriginal culture. In some ways, it reminded me of my own childhood growing up in Arizona among the American Indians.

After four days in Alice Springs, it was time for Helen and me to return to Cairns and our good friends Debbie and Nolan to fly to New Zealand on their continued vacation. Saying goodbyes on the patio would have been a sad moment had we not known we would be together again soon. Having returned to Cairns, it was time to turn our efforts to preparing *Tahlequah* for her next voyage, Darwin on the North West coast of Australia.

We had the benefit of consulting with a pilot familiar with the areas we were cruising known as the Gold Coast. Leaving Cairns with the *SV New Crusader* from South Africa, we did day sails stopping at Low Islet, Lizard Island, Flinders Island, Morris Island, Lloyd Bay, Margaret Bay, Adolphus Isla, and finally Red Island Point. Many of these islands are Aboriginal Sacred Sites and have significant meaning in their Dream Time. Entering caves, we came across many petroglyphs, magnificent works of art.

One evening, following a long passage, we anchored off Flinders Island, staying two days for the purpose of repairing our autopilot, which kept kicking out under a heavy following sea. The second day, repairs made, we dinghied across the Owens Channel to visit Stanley Island. We explored caves where Aborigines lived until the 1960s. We saw cave drawings commemorating the first European visitors by sail and documenting many of their medicinal plants.

While exploring the island, we neglected to consult the tide tables. Landing on the island at high tide and attempting to leave at low tide, our dinghies were now a third of a mile from the water. Our choices were limited; we could wait six hours for the water to rise or attempt to move the outboards and dinghy separately.

Brian attempted to lift his outboard; it was impossible. Michael threw the outboard over his shoulder and began walking towards the water. Sweating bullets, Michael managed to transport both outboards, an impossible feat for the rest of us.

That evening, Brian and I visited an anchored fishing boat to ask if we could purchase fish. Their response was *no, come back in two hours, and we'll have a barbie.*

We returned that evening with our wives, wine, beer, and snacks. The fisherman, his wife, and son explained how they caught fish, prepared it, and transferred it to the mother ship on a weekly basis. We partied all night, sharing sea stories and many beers. It was a wonderful evening that the women enjoyed as well as the men.

The day before transiting the Torres Straits at the Northeast Coast of Australia, we stopped at Red Island Point. It's a picturesque area and is considered Aboriginal Home Land. Although many white people live there, they're prohibited from purchasing land. The streets are red dirt and dusty; clearly we were now in the Australian Outback. The Regional Government Office volunteered to give us a tour of town and the surrounding area, and we enjoyed it greatly.

The following morning we transited the Torres Straits; there would be only one more stop at Clarence Cut just north of Darwin. We planned this stop due to the nine-knot current in the North Channel; it was important to pass through the North Channel at slack tide.

Several days later at 1:30 in the morning, I saw lights closing on our stern. It was a large fishing vessel moving at 20 knots. I radioed I was altering course 20 degrees to starboard, there was no response, and he appeared to be altering his course to mine. Continuing to close distance, I radioed him a second time, and he responded in broken English.

I requested he alter course immediately.

He said, "I get captain!"

By now, the vessel was continuing to close the distance between us.

Their captain came on the radio. "*Tahlequah,* I don't see you," he said.

"For God's sake, man, my running lights are dead ahead of you," I responded, adding, "I'm putting on my mast emergency strobe light."

Once the light was illuminated and flashing, he responded in an alarmed tone, "I'm altering course right away."

That was a close one, I thought, *either his radar wasn't working, or his mate was blind. Another minute and he would have hit me.*

During the trip to Darwin, we were frequently contacted by Australian Coastal Patrol aircraft, requesting we identify ourselves. We received instructions to proceed directly to the quarantine area at Coleen Bay Marina. Upon arrival, we requested permission to enter the Coleen Bay Marina but were told there was no room; instead, we were to anchor in Fanny Bay off the Darwin Sailing Club. We were requested to stay aboard until visited by Customs, Immigration, and the Department of Fisheries.

A boat full of divers inspected the bottom of our hull to validate there were no unwanted marine growth, mussels. Another man came to sanitize our through hull fittings. During this process, he removed a siphon vent, losing the small rubber grommet. After spending an hour looking for the part without success, I said, "Forget it. I'll buy a new one."

Although I didn't know it at the time, I was glad I was cooperative; he was also the lock operator at the Coleen Bay Marina.

After arriving at the marina, I received a SAT phone call from Safari. Ivor asked if I could help, he was informed by the Rally that the marina could not accommodate a boat of his size. Going to the marina office, I received the same response. Asking if there were alternatives, it was explained that half the dock spaces were privately owned, giving the marina control over them.

Pressing the need, I was provided contact numbers to negotiate my own arrangements; admittedly, the marina wanted nothing to do with it. I contacted a holding company that owned a slip previously used to berth the owner's private yacht measuring the same size as *Safari*. I explained the situation; he offered *Safari* his slip to be used for the duration of the Rally's stay in Darwin. It was explained to me later, not everyone was happy with this arrangement. I received a reprimand from the Rally representative for not informing the Rally of our private arrangement. Thanks to the cooperation of the slip owner and the marina, *Safari* was able to stay with the other Rally Yachts at Coleen Bay Marina.

Helen, Michael, and I along with Jill and Brian of *SV New Crusader* stayed at the Kakadu National Park for three days. We drove to Ubirr Rock to see the sunset, returning the following day to see the caves and petroglyphs. Some rock paintings dated back 5,000 years. Aboriginal peoples have been living in this area for 20,000 years. Some caves had drawings intended for learning purposes. I tried to imagine the many people, memories, and stories here. Climbing the cliffs was extremely difficult given the worsening condition of my hip; I used my hiking stick to assist in the climb.

We took a two-hour boat ride at Yellow River, and we saw many crocks, including a giant one in the wet grasslands. We slowed down for a boat in front of us, and as the boat turned, we saw many children sitting with their parents, appearing upset. When we replaced the boat in front of us, we saw the reason the children were upset. A very large alligator had a small wallaby in its mouth; it was trashing the animal from side to side against a tree. Apparently the children saw the gator catch the cute little wallaby, then proceed to kill it. So much for the innocence of children.

We loved Australia and wished we could stay longer. Being part of a Rally by nature implies schedules must be respected. Reducing the risks of bad weather throughout a circumnavigation requires we arrive and leave areas as scheduled. Like anything in life, there's an upside and a downside to everything we do. In our case, the benefits of being part of a Rally far outweigh the disadvantages of being on our own. Although having been able to visit more places on our own, we would have been away from family and friends for many years. This doesn't discount that some day we may do another circumnavigation.

CHAPTER 16

We returned to Darwin to re-provision and prepare for the next leg of our journey. We got our Indonesian Visas and Australian Port Clearance, and our next stop was Indonesia. Several days later, approaching Indonesia, we encountered hundreds of floating fishing platforms, constructed of bamboo poles lashed together with flags to warn passing vessels. As these were unlit, we maintained a careful lookout throughout the night for these navigational hazards. Indonesia consists of 17,000 islands, stretching more than 5,000 km between the Australia and the Asian continental main island dividing the Pacific and Indian Oceans on the Equator.

Indonesia has the world's largest Muslim population, needless to mention our concerns as an American Flagged Vessel. Our fears were soon relieved and best described through the following story. Short on fuel, *SV Ishtar* and *Tahlequah* stopped at an isolated island anchoring in the main harbor. Going ashore, Peter and I were met by a man offering to take us for fuel. The experience was a novel one; he took us to several old drums along the roadside. Opening the first lid, he tasted the fuel saying, *American Crude*. Not to be outdone, I dipped my fingers into the diesel, tasted it and said, *Yep, from California*. His wife dressed in Muslim clothing dipped a scoop with brass plate stating one liter into the diesel and filled our jerry cans. Back on the boat, I filtered the fuel through a Baja Filter; it was the cleanest I'd seen in a very long time.

Our host offered us a tour through town, which we accepted. Most taxis were horse-drawn carts, but only the best for us, a motorcycle with open cart. The streets were crowded, and driving through the crowds, he yelled, "I have Americans, and I have Australians. I have Americans, and I have Australians."

Thinking this might mean we were available to the highest bidder for hostage taking, I was aghast. Instead, we were amazed to see people smiling and waving to us. Children approached speaking English, welcoming us to Indonesia.

Being hot, we requested our driver stop for a drink. He reminded us it was the holy month of Ramadan, and nothing would be open. He took us to a closed shop requesting we remain in the cart, and he disappeared behind the building. Reappearing, he asked us to follow him to the back. Entering through the rear, a Muslim family offered us soda and food. We accepted soda and managed to communicate with our hosts in English. They took great interest in the US and places we visited in our travels. In our honor, they put an American rock video into their VCR for our entertainment.

After returning to the harbor, we were greeted by the Harbor Master, a young man in his twenties wearing an official uniform. After learning I was from the US, he invited us to anchor overnight in the harbor, stressing we were safe there, adding if we had problems to contact him on the radio. Reassured, we agreed to stay the evening. Early morning and late evening, we heard the call to prayers from local Mosques, a very beautiful, melodic, and moving experience. This story is indicative of Indonesia and the Muslim population here. The islands of Kupang, Bali, and Nongsa made us feel as special and honored guests. Everyone waves, attempts to communicate, and makes us feel welcome.

Indonesia has many lighted and unlighted fishing vessels, making our journey sometimes hazardous. Warned in advance of fishing vessels taking an interest in us, we were especially cautious. On several occasions, some came close to investigate. Following several experiences, we accepted their inquisitiveness as normal behavior. During day light hours, they waved and were friendly. Strong counter currents, shipping lanes, and fishnets kept us on constant alert, especially during night watch.

Our official entry into Indonesia was Kupang, another official Rally Stop. The Rally advance team facilitated our Customs and Immigration check-in. The mayor of Kupang treated us to an official welcome dinner, presenting each of us with a hand-woven traditional weaving. I wore mine over my shoulder for the duration of our stay.

During a several day stay in Kupang, we took a tour of the island. Visiting a local museum, we encountered school children. Both children and teachers were as interested in us as we were in them. Many wanted pictures taken with us, and teachers posed with us while the students flashed pictures. Each time we took a picture of the children, they screamed in excitement. We visited traditional villages throughout the island. Many villagers wore traditional dress and lived as their forefathers had. It was not uncommon here for a man to have several wives and many children. During a visit to one remote village, a woman offered her small baby to us. Needless to mention, the offer was declined, and we made a hasty retreat.

We were familiar with the Indonesian custom that if you admired or asked for something, you would likely receive it. The tour bus stopped for lunch at a restaurant located at a scenic overlook. Because there were so many of us, it took a long time to be served.

Carolyn (*SV Nademia*) admired the beautiful ceiling light fixtures made of wicker to resemble lanterns. She called the waitress to ask where they were purchased. The manager came to our table to tell Carolyn he purchased them many years ago on a trip to Malaysia.

Carolyn jokingly said, "Would you be willing to sell me one?"

The manager left the table to return to the kitchen. A few minutes later the waitress returned with a ladder, with a knife she cut the electrical cord to the lantern, lowered it, and presented it to Carolyn as a gift. We were amazed that anyone would honor such a request, let alone present it as a gift.

On October 8, we left for Rinca, one of the two islands where the Komodo Dragons live. The landscape looks like Jurassic Park, and no one lives here because it's a National Park, and unlike Komodo Island, few people come here to visit. We anchored in 100 feet of water and stayed two days.

During this period, the young crew from *Safari*, Franco, and William joined Michael to go spear fishing. Attempting to anchor nearby a reef, a diving boat began waving them off, screaming profanities. Moments before, Ivor had picked me up to take a tour of the bay. Returning, Helen was frantically pointing for us to return to *Safari*, indicating something was wrong. Bernice had locked herself in the cabin below and called Helen on a handheld radio when masked men boarded *Safari*.

Seeing a large powerboat tied to the stern of *Safari* with masked men carrying automatic weapons, we sped back to *Safari*. Arriving, we were greeted by men claiming to be from the Wildlife Management Services. Apparently, we were in a National Park, requiring a special permit we didn't have, and if that wasn't enough, we were attempting to spearfish illegally. We protested the permit requirement and emphasized the Mayor of Kupang assured us the Rally was welcome to visit without additional paperwork other than the general cruising permit we were issued.

We were informed we must pay $50 per boat for a permit to remain and no more spear fishing. *Safari* refused, saying he would leave today rather than pay $50.

Turning to me, the man in charge asked when I would be leaving.

"Tomorrow," I replied.

Without asking for money, he left with the promise we would not spearfish. At that very moment Franco, William, and Michael returned to the boat with spear guns in hand.

We saw several Komodo Dragons walking the beach. Although these giant monitor lizards appear docile, they move quickly, eat meat, and can easily devour a goat. Two days later, we sailed to Komodo Island, the most famous of the Komodo Dragon Islands.

We were informed we must purchase a five-dollar permit to anchor in a National Park, a far different amount than the $50 requested within the same National Park on Rinca Island. A Ranger on the island gave us a two-hour walking tour; we stood within 10 feet of a large Komodo Dragon. Fortunately, he had eaten and had little interest in us. Later, we explored beautiful white sandy beaches and swam over beautiful coral reefs. Local craftsmen came to our boat attempting to sell us hand-carved Komodo Dragons and other items. Following the usual price negotiations, we purchased several beautiful items to bring home as gifts.

Following a several-day sail, we arrived at the beautiful island of Bali. As the sun rose at 5:30, we discovered we were in the center of thousands of small sailboats with outriggers. Initially, I thought we ventured into the world's largest single design regatta. It was only when I saw the occupants fishing that I realized it wasn't a regatta. These single-crewed sailboats were unique and unlike any fishing boats we've seen before.

Bali is a beautiful island. We discovered each village specialized in one of many trades. These varied from stone carving, wood sculpture, painting, weaving, silversmith, and batik, (hand painted cloth). Quality was excellent and prices affordable; their skills were passed down from generation to generation within each family and village.

We visited a 960-year-old Hindu Temple consisting of many structures extending to the top of the mountain. It was necessary to take a motorcycle the last mile to reach the temple. We were required to purchase traditional clothes before entering the temple grounds. There were thousands of worshippers and priests offering prayers. A guide explained the basic beliefs of Hinduism, a remarkable religion. Reaching the highest temple, we were given a small bowl containing flowers, rice, and burning incense. A priest blessed us, and he pressed rice to our foreheads while putting flowers in our hair.

Helen and I stayed for several days in the town of Ubud, a unique town with many stores, selling weavings, carvings, paintings, etc. Hindu temples are everywhere in Bali, and each town has many beautiful temples, one larger than the next. We visited the Old Palace where we witnessed Balinese Dancing, telling a traditional story. Balinese dancing is not unlike Chinese Opera. A colorful dragon, monkeys, and other animals are a rich part of the story while young women dancers perform traditional dancing. The main town of Bali is modern with European, and Australian tourists, beautiful beaches, shops, and tourist sites. Bali is a place where Westerners can feel comfortable without sacrificing life's many comforts.

Visiting the site of the Bali nightclub bomb blast and Bali Peace Park across the street is both a sad, sobering, and inspiring experience. Loved ones, friends, and others continuously leave inspiring poems, writings, and symbols of this tragic loss on the fence. The many names and countries of those who tragically died on this site are inscribed on a marble monument across the street.

After reading the many personal messages written on the banner, each of us left silently fighting back tears.

Oct. 12, we arrived at Nongsa Point Marina for the purpose of checking out of Indonesia. Here we re-provisioned, swam in the marina pool and prepared for our short 23-mile journey to Singapore. On Oct. 17, we left for Singapore, sailing parallel to the shipping lanes then transiting the narrowest corridor. Singapore is the busiest port in the world; sailboats must exercise caution being ready to take evasive action if needed. Currents are swift and small craft move in and out of the shipping lanes without warning.

We arrived at the Republic (formerly Royal) of Singapore Yacht Club early afternoon and were assigned a berth. The Yacht Club, founded in 1826, is modern having many facilities, including excellent restaurants, hotel, pool, health club, chart room, conference room, Internet services, and repair facilities. The Yacht Club sponsors and participates in world-class racing and has a distinguished history.

Because we received a three-month membership, we enjoyed checking into the hotel for two weeks. The Rally support team greeted us on arrival and facilitated our Singapore check in. We immediately arranged to have a new step down transformer installed to convert 240 VAC to 110 VAC, allowing us to run our air conditioner. The Yacht Club treated us to several special events in honor of the Rally's arrival, including a dinner.

Although I've been to Singapore many times in the past on business, I no longer recognized this city. Singapore is a world-class city consisting of skyscrapers, Paris-like shopping, exquisite dining, entertainment, park, and sports facilities. Land reclamation continues at a frenzied pace with marsh and waterfront areas continuing to disappear. Infrastructure is outstanding and equals anywhere in the world. Visiting the Sim Lim Tower and Station areas, we discovered electronics at affordable prices. We purchased hardware to improve both navigation and storage of data. Singapore has some of the world's best shopping and something for everyone.

We celebrated Thanksgiving Day with our American friends, Dick and Leslie York (*SV Aragorn*) at the Inter Continental Hotel in downtown Singapore. It was a delightful dinner in true American style with turkey and trimmings. Celebrating holidays traveling abroad for extended periods helps maintain our connection with home. Following a Skipper's Meeting at the Republic of Singapore Yacht Club we prepared for the next leg of our journey to Phuket, Thailand, where we remained through Christmas and the New Year.

CHAPTER 17

Leaving Singapore, heading to Phuket, Thailand, we made several stops, including Port Dickson, Pangor, Paya Island, and Langkawi. From the time we left Singapore, we stayed to the Eastern edge of the shipping lanes to avoid the risk of collision. A 120 miles north of Singapore the shipping lanes end, requiring us to keep a sharp eye for ships and fishing vessels.

The island of Paya is a beautiful one, having fine white sandy beaches and palm trees. Many tourists come here to snorkel and dive. While snorkeling above a reef, Helen, Mike, and I saw several sharks swimming nearby; fortunately they took little interest in us. Leaving Paya en route to Langkawi only 20 nautical miles away, we encountered two whales off our starboard side. They followed us into the Langkawi harbor, then went about whatever whales do when they get bored following sailboats. Whales are a majestic sight as they surface spouting water high into the air.

We arrived at the Langkawi Yacht Club at 11:00 am in time to go into town to re-provision ship's stores, while Michael jerry canned more fuel back to *Tahlequah*. We always found time to visit the duty-free shop to purchase gin and wine. This evening, we had dinner in the Yacht Club with our friends and crews from *Safari* and *New Crusader*, both South African yachts. We always include crews when getting together.

The next day, more Rally boats arrived, giving us an excuse to enjoy another dinner at the Yacht Club with *SV Saint Barbara*, *SV New Crusader*, *SV Condor*, and *SV Nademia*. After many beers, we began talking about the type of people we are to put ourselves through the hardships of circumnavigating. This naturally led to the next topic, the hardships on our wives. I learned something this day: don't drink before having this discussion, or leave your wife on the boat.

Helen, Michael, and I took the opportunity to visit the underwater aquarium and sky ride to the top of the island's highest peak. The panoramic view was breathtaking but a little high for the likes of the captain. The next morning, we officially checked out, obtaining Port Clearance. We took one last swim in the pool to refresh ourselves and retired to be ready for an early start.

Leaving at 6:00 am, we found the winds light and the seas calm, and it was necessary to motor sail. After dark, we began to see many lights on the horizon. Within hours, we found ourselves in the center of groups of fishing vessels. Each time we passed one group, another suddenly materialized. The radar proved useless, and we finally abandoned trying to use it. Michael was stationed on the bow with a night vision scope, picking up those vessels without lights. The entire night, we spiraled our way through a sea of vessels until sunrise.

We appreciate when fishing vessels see us; they make every attempt to alter course. Unfortunately, many are occupied fishing; it's therefore up to us to avoid them. On the sea. a fishing vessel has right of way over sailboats. Not willing to take chances, our crew discusses course changes before they're made, a system that works well for us.

Traveling in the company of *SV Paroo*, I observed a large vessel moving towards us on radar. Although it was some distance away, its speed was 26 knots. I radioed *Paroo* to make them aware of the situation. As the vessel came within eight nautical miles of us, I knew the burden was upon us to alter course. Believing it was safer to fall off to starboard, I radioed *Paroo*. Dave informed me he was falling off to port and did so immediately. Not wanting to force the ship between us, I made the quick decision to fall off with *Paroo*.

Because the air was light, we started our engine to increase the safety margin between us and the approaching vessel. Without warning, the engine sputtered and died, placing us in the direct path of the oncoming vessel. Remaining at the helm, I asked Mike to bleed the injectors, hoping this was the problem. The engine restarted only to sputter and stop once again. The vessel continued to approach us at 26 knots. I radioed the vessel hoping they would alter course to avoid us, but there was no response.

I asked Michael to continue to bleed the injectors in an effort to get distance. In addition to our tri-color mast light, I illuminated the deck running lights and emergency strobe light. Each time the engine started, I throttled the engine hoping to get a few more meters' distance. During this same period, I frantically attempted to reach the oncoming vessel using my cockpit VHF radio. It appeared the vessel began to slow, but it was too late for them to alter course. The vessel was now on top of us, and I had one last opportunity to gun the diesel before it stopped. As the vessel passed, I looked up to see the faces of the officers on watch looking down at us from the bridge. Their faces matched my own look of terror. The clearance between us and the oncoming vessel could be measured in meters.

A rule aboard *Tahlequah* is "no drinking during passage." After the ship passed, I asked Michael to get me a strong whiskey; I needed it. I learned one important lesson this day: make the decision right for you.

Knowing we would arrive in Phuket in a few hours, we showered and dressed. Thailand is a Buddhist country with a population exceeding 60 million people. The Thai people are renowned for their friendliness and hospitality, and the country is commonly referred to as The Land of Smiles. A favorite saying in Thailand is: *no money, no problem.*

Thailand has a sophisticated infrastructure, including hospitals, schools, shopping, military, and police. Until recently, Thailand has been the model for reducing HIV in Asia; unfortunately, this has changed due to reduced spending in educating youth. Thai women are beautiful.

At first light we saw Phuket, Thailand, a beautiful scene with pillar-like structures spiraling skyward from the turquoise clear waters. I loved Thailand; I was returning home. I had been here many times over the years on business, had many friends here, and could think of no better place I wanted to be for the Christmas holidays. A few hours later, we arrived at the Yacht Haven Marina in Phuket, and we were assigned a slip especially reserved for Rally Boats. After official check in with Immigration and Customs, we went about our customary procedures of cleaning the boat and the most important thing of all, taking nice warm showers.

Because this was Michael's first time to Thailand, we were intent on showing him as much as possible. Knowing we would have to do these things quickly before Michael discovered the beautiful women of Thailand, we went about renting a car with driver to take our first tour. At a Thai show, we saw traditional dancers, kickboxing, and a martial arts exhibition. Later, we saw a monkey show where they retrieved coconuts from a tree. I felt sad that monkeys were being taught tricks to entertain tourists and wondered if they wouldn't be happier in the wild. We rode elephants in the wild, and at the end of the day visited a beautiful, newly restored Buddhist temple. On the third level of the temple was a tastefully constructed glass enclosure to house Buddha's tooth, where people came to kneel and pray.

In past years coming to Thailand on business, I frequently arranged back-to-back weekends in Bangkok, many times visiting Buddhist temples throughout the city. Here, I watched devout people coming to pray for hours, both young and old. Although I'm not a religious person, I admired the flexibility and tolerance of Buddhism and its effect upon the faithful. I've tied a fisherman's ribbon on *Tahlequah*'s bowsprit for good luck as do Thai fisher*men* on their Long Tail canoes. It's made of three colorful (red, green, and yellow) separate ribbons woven together.

Mike met young Thai women; an evening rarely passed without his having a date. Mike met a very beautiful nurse working at the Phuket International Hospital that he saw regularly. So much for showing Michael Thailand! I'm glad we showed him a few things while we could. The Rally arranged an official welcoming party for us at the marina. This included a buffet dinner with

Thai traditional dancing. We took a picture of Mike with his arms around four beautiful young Thai women.

We arranged with Carolyn and Allister from the *SV Nademia* (British Registry) to fly to Bangkok, spend a few days, and then fly to Chang Mai in the mountains. We stayed at the Royal Hotel in the old part of the city near the Grand Palace. We took Carolyn and Alistair to many of the places I enjoyed visiting over the years. These included the river market, the Temple of the Reclining Buddha, the Grand Palace, and the Marble Temple. My hip and back got so bad while touring the Marble Temple, I excused myself and left in pain to go to the hospital. Within minutes of arriving at the Spinal Clinic, x-rays were taken and a doctor came to discuss my prognosis.

Seems I had arthritis in my lower spine in addition to having it in my hip. After much discussion, he emphasized this didn't have to be a restrictive condition and would come and go. He believed I'd managed to pinch a nerve in my spine when lifting something. This, he reassured me, would go away. After being given pills for pain, Helen and I returned to the hotel.

Next we flew to the city of Chang Mai and stayed at the Chang Mai Holiday Hotel. This we did only after discovering the hotel we made reservations at was below our standards. I believe the rats were disappointed we left. Our driver for the next three days in Chang Mai arranged an informal itinerary for us. The first evening he asked if we wished to go to the Culture Center for dinner and a dance performance. We declined, saying it was late and instead would like to have dinner at a local restaurant and retire early. He agreed and offered to pick us up at 2000 hours in front of the hotel.

At the appointed time, he came to the hotel and took us directly to the Cultural Center for dinner and a dance performance. In the future, when asked what we wanted to do, we responded, *whatever you want*. After the show, we realized he was right, and it would have been unfortunate had we missed the performance. He was entertaining and fun, and he took great pride and pleasure in his job as our guide for the remainder of our stay in Chang Mai.

One day we did a self-directed walking tour, visiting five different Buddhist temples. At one temple, a woman sold us a small cage with little birds inside. She invited us to release the birds in front of the temple for good luck. It was a pleasant feeling to release the birds, watching them fly over the temple roof. We couldn't help but wonder if they were trained to return to be sold again. Later in the afternoon, we enjoyed lunch in a rustic restaurant overlooking the city. Each evening, we managed to take Tuk Tuk rides, one evening to the Night Street Market. We visited a silk factory, watching the actual making of silk thread from cocoons. The handmade weavings were beautiful and affordable.

We find ourselves purchasing beautiful items for family and friends at home. Once home, we rarely remembered which item was for whom.

We see child labor in every country, and it's not uncommon to see eleven- and twelve-year-olds helping shopkeepers, weavers, jewelers, and carvers. I've seen no evidence of physically abused children or children working in poor conditions. By western standards, seeing children working is unacceptable. Having traveled many years throughout Asia and the Third World, I view this differently. I've seen children in India starving to death in the streets. I prefer to see children working, contributing to their family's welfare.

We returned to Phuket to get more supplies and provisions for Christmas Day. The Rally was our extended family, and being together for Christmas was very important. The Rally organizers suggested Phi Phi Don Island as one place to gather for the holidays. The Sailing Vessels, *Aragorn* and *Sint Barbara* previously sailed to Phi Phi Don Island to see if a Christmas buffet luncheon could be arranged. Arrangements were made; we would sail to Phi Phi Don Island to celebrate Christmas together.

The day before Christmas, we made the five-hour sail from Phuket to Phi Phi Don Island. We decided to anchor on the north shore because unlike the south shore, it was peaceful and quiet. The many hotels, stores, shops, and numerous restaurants were invisible to us from this side of the island. Going ashore the day we arrived, we were amazed to discover how many people were here. This was the playground of mega-models, movie stars, and those that could afford to be here at this time of the year. It was Christmas; the hotels were filled with people from all over the world celebrating the holidays in a perfect setting. Before returning to *Tahlequah*, we had sundowners at the Jungle Bar Restaurant on the beach. The Jungle Bar was where we would have our Christmas luncheon on Christmas Day.

Christmas morning, we awoke and had a special breakfast Helen prepared for us before exchanging customary presents. We put on clean and pressed clothes before going to the beach. Many had already arrived, the bar was full, and most were wearing more clothes than I'd seen in a long time. We had a wonderful traditional Christmas lunch together. Following our luncheon Santa, (Alistair from *SV Nademia*) dressed in Santa's special extra-large snowflake earrings, presented each of us with a special Christmas gift. He was assisted by his wife, Mrs. Claus (Lady Carolyn, as we fondly call her). Later that evening, we had a dinghy drift, shared munchies, and watched another beautiful sunset—the perfect way to end a special day.

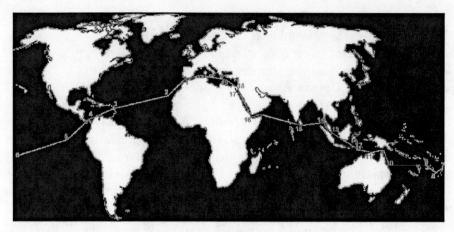

2003–2005
The round the world Blue Water Rally

San Blass Islands—Kuna Indian children in log canoe.

Transiting the Panama Canal—The Gaton Lock

Galapagos

Nuku Hiva, Marquesas—Ed's first traditional tattoo

Ua Pou, Marquesas—family going to church

Bora Bora

Ed, Hele, and Peter (a real Aussie)

Fiji—Captains wearing Sulus

Musket Cove, Fiji—Benno's & Margit's wedding

Ambrym Island, Vanuatu—children playing "Find the Fire Game"

Alice Springs (Outback), Australia

Australia—Debbie feeding Kangaroos

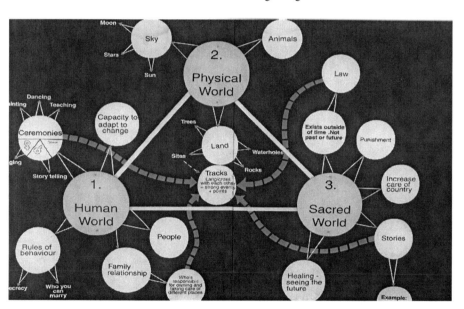

Aboraginal Dreamtime

Komodo Island—Komodo Dragon

Chang Mai, Thailand—Helen on elephant trunk

CHAPTER 18
Tsunami at Phi Phi Don Island 2004

Eight Bells

Those who love and work by the sea
The sea takes its own

Those who lament and pray by the sea
The sea takes its own

Remember those who perished in the sea
God's Sea knows its own

No words can explain or make sense of what happened at 10:45 am, December 26, 2004. The poem above is dedicated to those who perished at Phi Phi Don Island on that fateful day.

The day after this tragedy, my friend Dick York of the *SV Aragorn* came to the hospital to visit Helen and me. Asking what happened, I told him only of the people saved. I was hiding from the horror of what happened at Phi Phi Don Island, as it was too soon and painful to relive. Dick wrote my story as told, sending it to *Sail Magazine* for posting on their website. After reading the article, I felt compelled on behalf of the people who died there to write the following letter, sending it to everyone in the Blue Water Rally and its organizers. The letter below accurately describes what really happened at Phi Phi Don Island on the morning of December 26, 2004. This letter later appeared in magazines and newspapers throughout the world.

As part of the Blue Water Rally (A group of 20 sailboats from around the world circumnavigating together), 12 Rally boats gathered at Phi Phi Don Island positioned 15 miles off Phuket, Thailand, to celebrate Christmas day together. On the morning of December 26, Helen and I awoke and decided to go to the island for breakfast, leaving our grandson Michael sleeping aboard Tahlequah (our Sailing Ketch). It's important to explain Phi Phi Don is a small island with a beautiful beach on the north side and the south side having many open restaurants, numerous shops, and several luxury hotels catering to the hundreds and hundreds of tourists visiting each day. It's the perfect picturesque island vacation get-away, and it's world famous as a popular Thailand vacation area. Each side of the island is connected by a narrow walkway.

Strolling the boardwalk, we favored a small bakery with tables, enjoying breakfast together. This day, I saw many small children in carriages, babies in back slings, and the usual teenagers and families on Phi Phi Don. I took special note of peoples' faces and accents giving special attention to Americans. Ferries were arriving with hundreds of tourists emerging to enjoy Phi Phi Don for a day's visit. Following breakfast, we made a last-minute decision to stop at one of many Internet cafes to respond to e-mails from family and friends. In each e-mail we stated, "Wish you were here."

At approximately 10:45, we returned to the south beach to collect our inflatable and return to Tahlequah. Arriving at the beach, we were stunned to discover little water left in the anchorage, a phenomenon we're not used to. Helen remarked she thought Tahlequah might be on sand, I added this was impossible as we were anchored in 40 feet of water a short time before. We began dragging our inflatable through the sand to reach water in the distance. I saw rental powerboats and Long Tails (Thai Canoes) racing toward us skidding frantically but unable to make progress because of the sand. I commented to Helen how people abuse boats and how furious it made me. The skippers of the Thai Canoes motioned us back and began jumping from their canoes to anchor them in the sand.

Looking into the distance, I saw a small foam line on the horizon moving toward us. Helen and I agreed to abandon the dinghy and run back to the island beach for safety. Running, I continued to look behind to see the wave gaining distance at an unbelievable rate. Seconds later, I turned again to see the wave hit a rental powerboat; it broke apart as it fell in the surf. I realized it was useless to run. I told Helen to stop, and I bear hugged her. I remember saying to myself, "I'm going to forget I have to concentrate on hugging her; I can't release her no matter what."

We saw a boiling froth of sea coming at us with an increasing loud swishing noise; it seemed to go on forever. Foolishly, I dug my feet into the sand hoping to withstand the wave. As it hit, I felt us smacked to the sand instantly, and as we hugged, I could feel us tumbling like toy dolls head over heels along the bottom. The pressure and force of the water prevented us from surfacing. As my hands were ripped from embracing Helen, we both surfaced against two palm trees and were held there by the ferocious current. Helen was in shock, staring towards the ocean motionless. I held her repeating again and again, "It's over, it's over, we survived, and you'll be ok."

At that exact moment, we were hit by a much larger wave. I felt the palm trees give way, and again we tumbled together along the bottom rolling over the island. I continued to focus on not releasing Helen. I kept thinking we were going to hit something; we have to, and I waited for that moment. We continued tumbling seemingly forever; I was running out of air and knew I had to make it to the surface. Forcing us to the top, I had time to gulp a quick breath before being forced down again. When surfacing, I saw I was passing through the palm trees on the south side of the island and knew we were now going out to sea. Desperate, I had to make it to the surface again and made a final effort to reach the top.

I tried to surface but couldn't because of debris everywhere. I lost grasp of Helen a second time. My hand grabbed a floating cushion, pulling myself to the surface only to be forced below again and again. Swallowing water, I knew the end was near and felt death all around me. I remember feeling a sense of peace I had never felt before; everything seemed to go into slow motion, quiet and very peaceful. I recall saying to myself, "I wonder how long it takes to drown, and I wonder if it will be quick? It's over now, and it's ok."

My hand seemed to touch something rigid; it was a pipe. Grasping it, I pulled myself to the surface and saw Helen's head below. Grabbing her neck, I raised her above the water for air. She was unconscious, pure white, and just staring expressionless. Slapping her face, I kept screaming, "Keep breathing! Keep breathing! We can make it!" over and over again. I suddenly realized I had grabbed the long propeller shaft of a Long Tail Thai canoe. Helen slipped and began to sink below the water expressionless, and her face seemed resigned as if to say, "I've had enough let me go." Grabbing her chin, I raised her again, planting her chin into the metal framework of the adjacent upturned boat.

There was a man standing in the boat staring outwards, I screamed to him to help Helen into the boat. Looking, he seemed unable to move and continued staring. Attempting to raise myself into the canoe, I continued screaming at him, demanding he help raise Helen. Suddenly he reached out, grabbing Helen's hand, helping to raise her into the boat. Although lifeless, I knew she continued shallow breathing.

With all my strength, I pulled myself into the canoe next to Helen. She laid over whispering to me that she couldn't breathe and had awful pain in her chest. I continued to scream to her to keep breathing, that she had made it and everything would be ok; all she had to do was continue breathing. Twenty feet away, I saw a man raising a young naked woman into the canoe; I knew she was dead, drowned people were everywhere floating past us. Looking towards the island, I couldn't believe the destruction; all the hotels had collapsed and were sliding into the sea.

I saw two large wooden upturned fishing boats moving towards us caused by the rushing waters. Fearing if we didn't move quickly, we'd be crushed, I got the captain to start the engine to move to a safer area and into deeper water clear of the wreckage. Continuing to hold Helen in my arms and comforting her, I began to hear pleas for help coming from the water. People were clinging to whatever they could with what little remaining strength they had. One woman begged me to help her and grabbed a board to stretch it towards me in an effort to reach us. Reaching out, I realized it wasn't long enough and gave up. Looking down into the water, I knew the only way of helping her was to jump back into the water but feared I didn't have the strength left to get back into the canoe a second time. Afraid of losing Helen, my hands were frozen cradling her in my arms. I kept screaming to the woman that it was over and to just hang on.

"It will be ok," I said. "You're safe."

A young man suddenly grabbed the side of our boat trying to pull himself up. The captain placed a ladder over the side for him to climb up; he tried but didn't have the strength. I saw the captain finally move toward him but couldn't lift him alone. Putting Helen down, I helped bring him aboard. The captain then found a rope to throw a woman nearby a line, and he dragged her aboard. I insisted the captain had to move, or we'd be crushed in minutes by the upturned vessels now only a short distance from us. I wanted to say all the strength in my body was gone, but it wasn't. I had to save Helen with what I had left and wouldn't leave her.

I knew Helen couldn't last long without oxygen, as her lungs were filled with water, and because of her pain, I didn't know if her ribs punctured a lung. Motoring towards the 15 remaining long tails, the screams faded as we distanced ourselves. Other long tails were moving with us, and every captain was standing silently and motionless staring at the island waiting for the next wave. There was total silence; no words were spoken between the boats. Everyone appeared in shock and could only stare ignoring the screams coming from the water.

I asked the captain if he was ok.

Looking at me, he responded that he lost his niece on the beach.

Looking seaward, I saw a wooden square-rigger racing towards the island. I told him we had to reach that ship and get Helen aboard for her to make it. Leaving the safety of the other long tails, he motored towards the square-rigger. As we approached amidships, I yelled to the Swedish captain that my wife needed to get to a hospital.

He responded, "My ship is your ship. Where do we take her?"

His crew immediately helped lift Helen aboard and laid her on the deck.

I requested oxygen, and he responded he had a tank; unfortunately, it didn't have enough to make the four-hour trip to Phuket. I used his radio to call the Rally fleet in the North Bay. They informed me Tahlequah was safe; most boats had broken anchor and were headed to deeper water bracing for the next wave. I was told Michael our crew had a bad gash on his arm and needed medical attention. Our boat raced towards the North Bay while a couple from another boat Regardless *raced toward us with oxygen and medical supplies. Midge was a nurse, gave us both injections for pain and administered the oxygen. They returned to their yacht to brace for the next wave. Another Rally boat brought Michael with a bandaged arm withand Jesse (a paramedic from* Gaultine II*) to travel with us to Phuket.*

We raced towards a major port in Phuket only to be informed by the Rally boat Aragorn *that the port had sustained major damage. The captain suggested going north until we found suitable anchorage. Three hours later, we transferred Helen to a hotel atop a hill and were immediately transported to the Phuket International Hospital. Upon arrival, we were rushed into the emergency room. Paperwork was put aside; they wanted to know only the patient's name and injury. Within minutes, x-rays indicated her lungs were filled with water but no broken bones or organ damage. I was warned by the doctor the body would slowly absorb the water but there was a high risk of serious pneumonia. She would have to be transferred to a private room for the night, then to the Intensive Care Unit.*

The following day Helen appeared worse, with more chest pain and pneumonia. The x-rays indicated her lungs were continuing to fill with fluid. Antibiotics and painkillers were administered every few hours, and within two additional days, the infection was under control, and Helen was transferred back to a private room. Each day, she continued to improve and was informed she would be discharged after a week in the hospital assuming all continued to progress well.

The faces and screams of the people we left behind in Phi Phi Don Bay continue to haunt me. I can never forget their screams and begging for my help. I find myself walking the crowded hospital corridors among camera crews looking for people I might recognize from that day; they are never here and I will never know what happened to them. Riding to the hospital one morning with people from the Blue Water Rally, we made an unexpected stop at the university for a medical student to volunteer. They informed me the university was providing counseling for victims of the tragedy. After sitting in the car for five minutes, I said I had to stay and walked into the building. I saw people sleeping on the floor, blankets and pillows everywhere.

Limping to the table, I said I needed to talk to someone about what happened and was provided a counselor within minutes. The most difficult thing I'd ever done was talk about that day in Phi Phi Don Bay, but talking seemed to help me. I cried and cried trying to find the words to describe that terrible day. Life suddenly seemed so different; my drive for pushing life to the extreme became numb and how I regard myself different. It was a day that changed many lives.
Ed Muesch

Blue Water Rally friends at Phi Phi Don that day are responsible for saving Helen's life and saving *Tahlequah*. They endangered themselves in the face of more waves to bring oxygen, medical supplies, and assistance to keep Helen alive and accompany us back to Phuket International Hospital. These same people returned to the island that evening to assist victims hurt in the worst disaster of the century. I stood in a Rally Briefing Meeting a few days later, with tears running down my face, as I tried to find the words to thank them. I will never know the Thai young man who unselfishly helped pull Helen from the water in his own grief, but I will never forget his face.

The American Embassy contacted us on the first day of this terrible tragedy and provided ongoing support throughout our ordeal. Embassy doctors, nurses, even a psychologist provided support by visiting many times at the hospital. We could never thank all the people who helped us, Thai and Westerner alike. The Embassy staff seemed shocked when Helen informed them we would continue our journey. Some Rally members assumed we would return home following this ordeal, and others understood.

Helen said we'd done this for the last two years, and it had to be finished.

Frustrated, anyone I attempted to talk to would say, "It's not your fault; don't blame yourself,"

I didn't know where to turn. I wasn't asking people to forgive me; the people who could do that were dead. All I wanted was just one person to hear what happened that day. In desperation, I told my friend Ivor what happened and how I'd left people in the water begging for my help.

Ivor said, "I will think about this all night, and tomorrow we will talk."

The next day Ivor returned and said compassionately, "I would feel the same as you, blaming myself I couldn't help more people."

Ivor's simple words helped me without lessening the truth of what really happened that day. A friend did for me what professional counselors couldn't.

I was still disoriented when the *NBC Nightline* came to the hospital; it was the fourth day after the tsunami, and I was still walking the halls looking for the faces of the people I left behind at Phi Phi Don Island. When initially asked to do the interview, I refused, but after a hallway discussion with the director, we saw the importance of raising funds for the victim's families and rebuilding Thailand.

When I asked Helen if she would be interviewed by NBC, she spruced up for the first time since the tsunami and said, "Where's my makeup!" I knew from that moment that Helen would recover.

We discussed hiring a Professional Delivery Skipper and Mate to take *Tahlequah* from Phuket, Thailand, to Djibouti while Helen convalesced. We interviewed several recommended skippers and were favorably impressed with Johnny "O" and his mate Serun. Their ship's papers and letters of recommendation spoke for themselves. We took *Tahlequah* for a half-day sail; undeniably *Tahlequah* was in safe hands until Djibouti. Johnny "O" is Indonesian, living in Thailand, working as a Delivery Skipper, and his First Mate Serun is Thai with equal experience.

Recovering at the Phuket International Hospital, Helen and I spent New Year's Eve 2005 looking from the room window. No New Year's Eve celebrations, fireworks, or parties were evident anywhere; this night was no different than another. Thailand remained in a state of shock as everyone was affected by this terrible tragedy; even the King's grandson died in the tsunami.

Before leaving the hospital, I received a phone call from *The Oprah Winfrey Show*. A woman director asked if they could do an interview. Helen loves Oprah and was thrilled at the prospect of having anything to do with Oprah and the show. They requested we fly to Chicago to appear on the show, but we refused as it was impossible for Helen to travel so soon. The director asked what our plans were after getting out of the hospital. I stated Helen and I discussed returning to Phi Phi Don Island for closure. She asked if their film crew could return with us, and they would cover the cost of our transportation. We agreed on condition it would be postponed until Helen was in the apartment and felt well enough.

The staff of the Phuket International Hospital made no distinction between Thai or Westerner; everyone received the same compassionate and caring treatment based upon injury. A doctor held my hand as he guided me through the hospital waiting room to the ICU where Helen stayed several days recovering. This simple compassionate gesture gave me the strength needed to get through the next several days.

Helen's lungs were now clear of infection, and x-rays showed the salt water had been absorbed into her body. Due to a separated Achilles tendon, she was limited to walking short distances using a cane. The doctor felt she was recovering and would be able to walk within the coming weeks. Michael, our crew and grandson, made the decision to return to the US to rejoin his family in Athens, Georgia. We credit and thank Michael who was aboard *Tahlequah* during the tsunami for saving her from foundering on the reefs, and we will miss Michael, but we wish him the best in his next adventure.

A reporter recently asked how this event changed our lives, and Helen spoke for the both of us, saying life is a gift; we have to live every minute to its fullest. Afterwards, Helen and I agreed from sharing the same experience neither of us feared death.

When I couldn't break through the surface of the water due to debris and out of air, I accepted I would die. I felt a peace I'd never known before, and it wasn't frightening or surreal, but in the simplest terms a tranquility and peaceful acceptance. From this one experience, Helen and I feel neither life nor death need to be feared.

Within a week of being discharged from the hospital, we returned to Phi Phi Don Island with Oprah's film crew, followed by an all-night satellite interview the following evening in Phuket. A part of me wanted to return hoping to discover this tragic event never happened, a part of me wanted to return to see the people I left behind, and a part of me wanted to see that the island had been rebuilt after the tsunami as a testament to the people that died there. Returning to Phi Phi Don Island, I couldn't believe the destruction: the hotels, restaurants, hundreds of shops, private homes, boats that were now gone or reduced to a pile of rubble. Thanks to the generosity of our friends Barry and Christiane Cager, we agreed to stay in their apartment at the marina to fully recover for 30 days and then fly to Djibouti and rejoin *Tahlequah*.

Volunteers were still combing debris looking for bodies. A large tree with Buddhist ribbons winding around its trunk was all that stood. Although not a religious person, I found myself standing there trying to understand how this could happen and how so many could have died. I found myself silently asking the people left behind for forgiveness.

The Indonesian Delivery Captain we hired to take *Tahlequah* from Phuket to Cochin, India, and then to Djibouti were now underway. Although we'd been unable to contact him, we understood he was reporting positions daily through the Rally Radio Net Controller. Although there was presently a cyclone between his position and Cochin, weather models indicated this system was to begin moving north, making his voyage a safe one. One yacht forward of his position had been dismasted, and another experienced a rig failure.

We didn't see the Oprah interview until returning to the US. The woman and director working with us were sensitive to our fragile condition and needs. We were given the impression she was as concerned for our well being as getting an interview. After seeing the video, I wrote to this woman thanking her for her professionalism and sensitivity to telling our story.

It was emotionally draining for Helen and me, and as a result, we declined further requests for interviews, explaining we needed to heal and prepare to move on with the Rally. I received several calls from an Australian company claiming to represent major networks wanting to send a film crew to capture our story, and again we declined.

A group of yachtsman living here in Phuket came together to form a local relief program, the Coco De Mer Tsunami Relief Appeal, targeting small villages lost in the system. These villagers lost family members, their homes, fishing boats, personal possessions and presently have no means of supporting themselves. The Appeal intends to help those most affected by the tsunami and those who typically fall between the cracks. The sailors who take responsibility for this effort are personally known to us.

Wanting to help the families of the people who perished, we explored ways of helping. We met with the organization's founder, Barry Cager. We decided the best way of helping was to put together a formal presentation to be given at yacht clubs and other organizations in an effort to raise money for this cause. Helen and I decided do this when we returned to the US for routine visits.

Our purpose for remaining in Phuket, Thailand, was achieved when Helen's doctor gave her a clean bill of health, including permission to continue our travels. Helen and I planned to be in Djibouti around February 10 to rejoin the Rally. Because I was unable to get a Djibouti Visa, Helen and I went to Bangkok one week in advance hoping to get one at the French Consulate there. In Bangkok, I was given an opportunity to spend a day with a Buddhist monk for the purpose of healing from the emotional trauma caused by the tsunami. My best hopes were realized, and it was the first time I'd slept through the night since the tsunami. Having someone to guide me through the importance of looking forward and up, not backward and down, and most importantly, the reasons why was his gift to me. An enlightened man, I found him to have meaningful insights I'd never thought of. Most importantly, I can move on with my life.

Helen attempted to visit the only woman monk in Thailand; unfortunately, the driver got lost after a half day of searching. Helen intends to visit her when we return to Thailand in the future. We visited Wat Po (Temple of the Reclining

Buddha) several times, and each time, we hired a guide (often past monks) to explain Buddhism to us. Without going into great detail, we were moved by the tolerance, wisdom, and gentleness of the Buddhist people in Thailand, a people practicing their beliefs. We found many similarities between Quakerism and Buddhism.

There's a Thai word I won't attempt to spell, but it translates as *A Pouring out of the Heart*. I truly believe these words stand for the essence of the Thai people, compassionate, peaceful, accepting, and always offering to share what they have. There is no need to fear walking out alone at night or losing your wallet or purse because it will likely be found and returned, for such is Thailand, *the land of smiles*.

Even now I seem to be unable to escape the memories of what happened at Phi Phi Don. I suspect these memories will be with and a part of me forever. I will learn to live and deal with them. After continued counseling, I guess I'm prepared to accept and live with the knowledge that a part of me feels I could have saved more lives and part of me accepts I had to choose between saving my wife Helen or others. Although we're not religious people in a traditional sense, it's changed our lives.

We received the e-mail below from our fifteen-year-old grandson, Chris Handy:

Dear Grandpa,
After seeing you guys on Oprah, I can tell that you still feel guilty about not being able to save more lives than you did. You should be proud of what you did because you saved the most important thing in your life right now, your wife. I also learned that many of the people saved from the water died of disease due to the lack of medical help in some areas. I am proud of you for saving grandma and you shouldn't feel guilty for a choice that tough.
Love,
Chris

The following correspondence was written by Helen to family and friends during her convalescence:

Dear family and close friends,
I wanted to let you all know how I am doing and what I am feeling. Ed has given you all the facts. I passed out three times in the minutes that we were

being tossed in the wave but do remember some of the feelings I experienced. First regret that I would be hurting you all by drowning and then just a complete peace of mind when I thought it was really over. The rest of my life is a gift and I will live it with purpose and joy.
Love,
Helen

Dear family and friends,
One of the things I will never forget is the kindness of strangers. First the long tail captain that pulled us out of the water and then the Swedish captain of the charter boat that took us on the four-hour journey to Phuket and the hospital. Phuket International Hospital is first rate, not just medically but for the care and warmth given freely by the entire staff–from physician to the cleaning staff. Nursing care was efficient and professional, but the nurses still found time to sit on my bed and hug me. You all know how much I believe in hugs.

Three of Mark's (our son) coworkers from the AIG office Bangkok, Phuket, and Hong Kong visited the hospital and kept in touch with him. Talking to you guys on the phone was very important to me. It means so much to Ed and me that many of our good friends, both personal and from work have reached out to us. Their concern and caring have helped us through this difficult time. We plan to leave Thailand on February 4th to continue the Rally. We have mixed emotions leaving Thailand and are considering returning in the future.
Love,
Helen

Dear family and friends,
The American Embassy called the hospital the first night, and we have had visits from at least three representatives to be sure we had everything we needed. (Boy, maybe Bill Clinton will come check on me!) AND how about those television interviews? The folks were so nice, and it was so important to talk about what happened and to share what was truly a miracle. Hope it didn't upset any of you.

Our plans now are to have a professional captain take Tahlequah to Djibouti where we will meet the rest of the rally boats on Feb. 11. We will rest here at Boat Lagoon and then will sail with our group up the Red Sea and finally to Crete. In late April we will leave the boat in Turkey and fly home for Ed's surgery and a long visit with you all. Ed offered to bring me home right away. But *it is truly important to ME that we finish what we started with our rally group. They are our second family, and we can be safe with them.*

Thanks for understanding and for your love and humor. We laughed ourselves silly about Paul's comments how it was my buoyancy that kept us afloat (all my life I have found it impossible to swim underwater). As long as we laugh together, everything will be alright.
Love you all,
Helen

The story below is an eyewitness account of the tsunami aftermath at Phi Don Island by Dick and Leslie York of the Sailing Vessel *Aragorn*, a Blue Water Rally Yacht located in the North Bay next to *Tahlequah* off Phi Don Island during the Tsunami. We are very thankful to Dick, Leslie, and Tom who assisted Michael in saving *Tahlequah*. The following day Dick York sailed *Tahlequah* back to Yacht Haven Marina, Phuket.

Disaster Story

Just before nightfall, we anchored Aragorn and Tahlequah (still under our command) in another open anchorage about 1.5 miles north of our prior one, and decided to see if we could provide help onshore. Sloane, Tom, Catherine and Dick went in with medical supplies and clothing. We stopped by the NZ yacht Wind Dancer, and picked up a doctor's wife and a friend (whose boat had just been lost to the tsunami in Langkawi) and many supplies. We met the doctor coming out in his dink, and split the load. As we got to the beach, we had to pull the dinks up. We noticed the water was receding very fast, and we all suspected another tsunami wave, what the doctor called another event. We threw the dinghy anchors out, grabbed the medical gear and ran for the beach. There we found we could not immediately climb to high ground as the spit was filled with debris, and it was difficult to move around on it.

We had to move east along the beach for 440 yards or so until we could move up onto the spit. Then we picked our way along, getting to higher ground after climbing by houses that had been torn in half. Saw four bodies that had been laid out and covered (realized that many healthy people on the island had been helping others since the event happened). After climbing up hill, we met with some people who knew of injured requiring treatment. Simple triage meant the person with two broken legs and lacerations took precedence over the other person with the same injuries but severe internal bleeding ... there was no ER there.

Sloane a second-year medical student volunteered to stay on the island and assist the doctor. Tom, Cat and I wanted to work our way back to the dink to see the boats safe for the night. As we approached the dink on the beach, we were wading in 4 inches of water for the last 400 yards or so. With 30 yards left, the water level suddenly flooded fast, and was chest high by the time we got to the dink, and it was drifting away! We jumped in, and were able to motor away. Another anomaly. I now think the tsunami was bouncing around the Indian Ocean and would occasionally send a surge into places it already had visited.

We stayed at anchor watch until 1:00 am (27 Dec) when a surge warning came in over the radio. Leslie got Aragorn's anchor up, with Catherine and Tom, and I slipped Tahlequah's cable (second anchor, already buoyed), and we went to sea. We milled around all night, but did not notice any different wave. By dawn, we re-anchored, picked up the buoyed anchor.

We quickly motored in to the beach to carry more water and clothing to pick up Sloane, whom we had told we would come for at first light. There we saw hundreds of people helping carry injured to the makeshift helicopter pad/medical station. We helped carry two, the second who had been cared for by Sloane and the NZ doctor. Carrying wounded to helicopters, which are landing almost on top of you, is like MASH, at the wrong side of the captor trip. The scene in daylight was more of a disaster: pieces of structures everywhere, two stories of hotel windows taken out, injured everywhere, and white-wrapped bodies being pushed out to sea to meet the 24-hour burial rules of Muslims. We hated to leave, but had the responsibilities of two boats and five people to get to Phuket that night. As it turns out, mobile thousands were being taken off by ferries at the same time on the other side of the sand spit.

We found out that many of the injuries began at the legs and moved up. That is, debris hit people's legs as the waves washed over, and then took them down. But broken and lacerated legs were the first injuries. The more severely injured had more debris hitting them farther up, until their whole bodies were smashed by wreckage. Of course this does not count those pulled out to sea to drown. There will never be an accurate tally of those killed ... many will be missing as their bodies are at sea.

Thailand is doing a good job of providing aid. Their forces seem to be mobilized, and Phi Don has been effectively evacuated. It used to have thousands of tourists on it every day plus several thousand (?) Thais to provide services. Some remain in the hills, but the tourist Mecca bazaar that was on the sand spit is no more ... totally washed over by the waves.

We are now trying to regroup. Emotionally, it is difficult, as we have seen the disaster first hand. We were sorry to leave Phi Phi Don, but it is nice to know that almost everyone injured was evacuated by now. We are trying to figure out where to give blood that they need. The local hospitals are jammed, but Ed reports they are very nice despite the pressures.

The rally is on hold for a bit. The next scheduled stop, Sri Lanka, is devastated, and Galle could not cope with us in a few weeks. The stop after that in the Maldives, well many of those islands disappeared, as they were only about 3 feet above sea level. We will try to figure out a new routing and get underway when we can, meanwhile getting our visas extended.

Will let you know more when we can.

Dick

The story below is an eyewitness account on Phi Phi Don Island hours after the disaster. Sloane York, the daughter of Dick and Leslie York, is a second-year medical student who returned to Phi Phi Don Island following the disaster to offer medical assistance to those injured under the supervision of another doctor.

Hello Everyone,

I am sending this to just about everyone in my address book as I sit here in Phuket, Thailand. As my parents are sailing around the world, my brother, sister, and I came to visit them in Thailand for the holidays. We were anchored in the north harbor at Phi Phi Don on Dec 26th when we looked up and saw the water practically drain from the bay. For a minute we thought it was an interesting phenomenon since it was supposed to be high tide, and then we saw the water start to swirl and boats began to move uncontrollably. My dad realized this was not a curious spectacle and we quickly got life jackets on and started to get the anchor up as the waves rolled in. We had a few break across our bow before we were lucky enough to get our anchor up and get out of the harbor. Many boats broke their anchors (one boat recorded they were going 14 knots in the water as it sped by them—they normally do 6–7 knots on a good day of sailing). The waves crashed along the shore covering the trees and beach. We had no idea what horrors occurred there until much later. My dad and brother were able to take the dinghy out after it calmed down to rescue two boats that were unmanned and being flung around. Amazingly, they held up very well.

Later that night, after spending the whole day a mile off of Phi Phi afraid of another tsunami, we took some medical supplies into Phi Phi. I met a doctor off another boat and was dropped off with him for the night to try to help some people.

I have never beheld such horrors in my life. As we walked around this formerly picturesque tourist town, every single building had been taken out. The

water went as high as three stories and the hotels on the side of the mountains were even affected. I will never forget the piles of rubbish around the island. The concrete structures barely held and everything else was gone completely. It was incredibly frightening from sea but was nothing compared to what the people on land experienced. I wish I could tell you all the stories of every person I met as we tried to treat major lacerations and other injuries with the bit of alcohol, bacitracin, and gauze we were able to get off our boats. People said they saw hotel rooms zoom past them as they were pushed by the water.
Sloane

Eleven months following the tsunami disaster, I received the e-mail below from Barry Cager of the Coco Der Mer Tsunami Appeal.

Dear Ed and Helen,
Today we have been to visit Arunee Naweewong and Niracha. If the sea stays calm, Nirachu is coming to visit the boat with Kanokwan (our 10-year-old sponsorship) after school at 1530. We now have a Thai boat boy who speaks enough English to be able to communicate with the girls.
The fisherman who helped you in Phi Phi Don is Niracha's uncle-Arunee's dead husband's younger brother! He recognized you immediately from the photos you sent for Niracha. It is a small world and so strange that you should be helping his sister-in-law's child. I will post you some photos of today.
Kind Regards,
Barry

P.S. The boat visit was a great success and we have promised to take them sailing in November.

I could never imagine I would someday know who the kind fisherman was that helped save Helen's life. The odds of this person being Niracha's uncle (Prakoob Padungpol) are even more astounding.

Phi Phi Don Island, Phuket—Christmas luncheon

Phi Phi Don Island—3000 persons died

Phi Phi Don Island—destruction everywhere

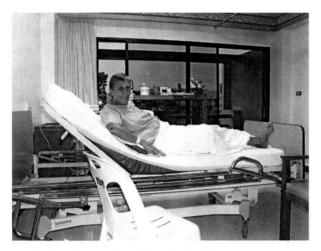

Phuket International Hospital—Helen recovering

Niracha—the little girl that lost her father
in the Tsunami—the future of Thailand

Michael returns home after the tsunami

CHAPTER 19

I've received several SAT calls from Johnny 'O' aboard *Tahlequah* indicating all was going well. Recent e-mails from Rally Directors confirmed the same. Although some Rally yachts were low on fuel, he used a spinnaker to conserve as much fuel as possible. The area from Cochin to Djibouti is considered dangerous. For this reason, the navies of bordering countries are aware of the fleet's daily positions. The Rally is broken into three groups of several boats, based upon speed. Prearranged coordinates are provided by Rally Control for each group. From these locations, boats maintain close proximity at an agreed-upon speed through Djibouti for security purposes. Wow! What bad luck, Helen and I will have to miss that one.

We left Bangkok without getting a Djibouti Visa; we were advised to visit the French Embassy in Nairobi, Kenya. The reason being this Embassy was experienced in acquiring Djibouti Visas and would assist us. Rally Control agreed to fax the French Embassy in Nairobi requesting their cooperation.

In Nairobi we went to the French Embassy and were told re-starting the process through them would guarantee Visas within 24 hours. Grateful for their solving our problem, we had no concerns paying the $100 fee a second time. Learning of our experience and need to rejoin *Tahlequah*, they did everything possible to accommodate us. The following morning, we were issued visas to enter Djibouti. Our experiences with the French have always been positive.

We received e-mails from Rally Yachts informing us Johnny "O" provided assistance to other yachts in the fleet. After the yachts re-grouped and fueled in Cochin, India, the Rally Directors wrote to inform us *Tahlequah* was in good hands. It was clear we had chosen the right Delivery Skipper. We enjoyed the story of a fishing boat coming alongside asking for food and liquor (pirates?).

Johnny "O" emerged from working on the alternator covered in grease and informed his intruders, "I'm hungry, too, but have to work, so PISS OFF!"

They responded, "Oh, Ok!" and promptly left.

Johnny "O" once took jerry cans into town for fuel, first removing his shoes. I didn't understand the significance of this until people began begging. He complained he couldn't afford shoes (meaning working for me), and they had more than him, and it always worked. I wondered if he'd ever received donations from beggars.

In Nairobi we stayed at the Hilton, it was a hotel from the early 1940s complete with furnishings of the period. Even bellhops were dressed as I recall them from my early childhood visiting New York. The lobby coffee shop was called

The Casablanca and had pictures on the walls of Humphrey Bogart. The service was outstanding, and we loved it!

Stopping at a small convenience store to purchase water, we were approached by a young woman carrying a small baby. Asking for change to feed her baby, we found ourselves suddenly surrounded by many angry women. They told us there was no reason a healthy woman like her couldn't find work on a farm and not to encourage this practice by giving her money. Not knowing how to handle this situation, we got into our van and returned to the hotel. We saw no other signs of begging in Kenya.

Having one extra day in Nairobi, Kenya, we hired a guide to drive us to the Nakura Game Preserve in the bush (a two-hour drive from Nairobi). The park had rhinos, zebras, lions, giraffes, warthogs, impalas, hippos, water buffalo, baboons, waterbucks, and thousands of flamingoes and pelicans on a beautiful lake. We had lunch in a park lodge where we observed the animals during midday. It was a wonderful experience that gave us a sense of Africa. We found the people of Kenya to be friendly, and helpful. Anyone attempting to sell us crafts seemed as interested in us as we were in them. They are a tall and handsome people; many of them wore suit jackets even in the bush.

We had an open discussion about AIDS with our driver and guide. He explained many families in Kenya refused to discuss this subject with their children and the need to protect them from disease. He revealed he spoke openly with his daughters about this subject and felt they understood the importance of protection. Clearly, he was proud of his children and his relationship as a parent with them. We were impressed with his presence and understanding.

Everyone seemed interested in telling us about the people and landscape of Kenya. Markets along the roadside reminded us of scenes from *Raiders of the Lost Ark* and *Out of Africa*. We were pleased to see government boarding schools located everywhere, especially in the countryside.

We noticed a small Quaker church along the road; our driver told us there is a big Quaker Meeting in Nairobi. Hearing someone inside, we entered; a man sitting by himself was Quaking (praying out loud). Reluctant to interfere with his meditation, we left after waiting a short time. This was a traditional Programmed Quaker Meeting (Evangelical), unlike our own Un-programmed Meeting (Liberal) back in the US. Regardless, we would have appreciated talking with him and making a small contribution. We loved Kenya and its people, and we look forward to some day returning.

We arrived in Djibouti on February 10 to await *Tahlequah*'s arrival. Djibouti was once a French Colony and was granted its independence in the 1970s. There is high unemployment and limited infrastructure. Many sleep along the waterfront hoping for opportunities to load and unload ships upon arrival. Our introduction to Djibouti began leaving the airport; several men took our luggage and began toting it off to the taxi area. Two men attempted to carry the same bag and got into an altercation when each attempted to put it into different cabs; for a brief moment, I thought they would tear our luggage apart. Fortunately for us, the Sheraton sent someone to meet us and mediated their dispute.

We stayed at the Sheraton Hotel awaiting *Tahlequah*'s arrival. The hotel was filled with German officers and French Foreign Legionnaires. The French Foreign Legionnaires looked like UPS men in shorts with Beau Geste hats; they were impressive. It was easy to tell who was who around the pool; the Legionnaires appeared to be lifting weights since birth.

Riding in a cab, we first agreed on cost; the driver always got lost in spite of our providing a map with specific directions. On arrival, we were informed it cost more, because it took longer than expected. Using our debit card to withdraw cash at the bank became an all-day ordeal; local people accepted this as a normal routine.

While awaiting the Port Captain at his office, Peter and I decided to cross the street to have a Coke. Coke open in hand, we spotted the returning Port Captain and ran out to make contact. Behind us people were screaming; maybe someone robbed the restaurant, we thought. Apparently we weren't permitted to leave with the bottles and were forced to return. The manager took our Cokes, informing us he would put them back in the cooler until we returned, and we agreed. However, when we returned, he knew nothing about our arrangement and offered to sell us new Cokes. We left without argument but with much amazement. Few tourists come to Djibouti; most are connected to the military alliance.

The Djibouti stories go on and on, and our reason for mentioning them is because we've discovered our impressions of a country are sometimes the result of our own limitations. Many Rally participants took a tour outside Djibouti, visiting the salt flats and nearby countryside. Their impressions of the country and people are more rounded and favorable than ours. We declined the opportunity to tour the countryside because we wanted Helen's leg to completely recover before touring Egypt next month. We've learned that initial impressions aren't always fact and seeing more of a country puts you in touch with both people and culture.

The Rally was given permission to anchor in a naval military zone and would be protected by Djibouti Naval security. There were many French, German, U.S. and Italian war ships. We were informed if we had any mechanical or electrical problems, the French Navy would assist us. Because of alternator problems, I requested assistance. I was invited aboard a French war ship, greeted by the captain, and informed they would assign someone to help solve our problem. Learning I was an American, he seemed eager to help in any way possible; needless to mention, we were impressed.

Knowing *Tahlequah* would arrive after dark, we planned to greet her at first light and were given special passes to enter the military base. Seeing *Tahlequah* again brought tears to our eyes, and it felt so good to be back aboard. Because the boat had to be cleaned, we remained in the hotel three additional days until we could get the boat in good condition.

We made the decision in Djibouti to retain our professional crew, Johnny "O" and Serun, through the Red Sea to Egypt. This would give Helen the opportunity to rest and regain her strength without the burden of three-hour watches in rough weather. Our arrangement was I would do day watches, and Johnny "O" and Serun would do night watches. The Red Sea is known for high winds and rough seas. Within 12 hours of leaving Djibouti, the winds increased to 55 knots and seas to 16 feet, and it was the roughest weather we've experienced anywhere. After 24 hours of bashing into wind and seas, a decision was made to seek refuge at an island off Yemen to wait for calmer conditions. Two boats forward of us made the same decision, and we used their anchor lights as navigational aids to enter a secluded bay along the Yemen coast in the dark.

Approaching the island, we started the engine, and the propeller shaft separated from the transmission causing a loss of power. After three attempts to sail into the bay, repeatedly the wind shifted, making entry impossible. In a final effort, our rudder jammed preventing steerage; the wind had now increased to 70 knots. Reducing sails to maintain stability, we pointed high to prevent taking us too far from the island. Another American Rally Yacht, *Aragorn*, came to our aid and attempted to tow us back to the anchorage. Instead, we found ourselves in the unexpected situation of towing *Aragorn* due to overpowering them. Because of the high wind and seas, it proved impossible, they returned to the anchorage alone. We agreed to remain off the island all night, work on freeing the rudder, and requested another yacht with twin diesels to tow us in the morning. If this failed, we would sail directly to Egypt.

Tahlequah, because of her low freeboard and ballast, behaved well, even in these extreme conditions. Although taking waves over the boat at no time were we in danger or in a situation we couldn't handle. Nevertheless, we needed to find a protected anchorage where we could make the necessary repairs to the engine. During the night, we freed the rudder allowing us steerage. Johnny "O" was confident he could repair the gearbox in a protected anchorage. The alternative was to continue sailing another nine hundred miles having no chance of repairing the gearbox. The following morning the rudder was free, and *Paroo* came to our position to tow us. We hoisted a main sail and were towed to the Hamish Island Group (Yemen) 10 miles away. We were aware that yachts were not welcome here without advance approval and hoped for sympathy.

Approaching the shoreline, a military powerboat swept down upon us with automatic weapons circling us several times. Seeing our towline, they waved and directed us towards the only anchorage along the northwestern shore. They were genuinely friendly and seemed to want to help. Entering the secluded anchorage, we didn't see them again until the following day. Several Naval Officers visited us aboard a chase boat and asked what the problem was. We informed them repairs were completed but winds remained at 50 knots, and we needed to remain until weather conditions improved. To our delight, they said, *Fine, long as you need and we're off.*

The anchorage was beautiful with white sandy beaches and light blue water along the shoreline. There were high cascading mountains with very old stone houses built on top. The steep mountains looked like the result of a lava flow, didn't support vegetation, and were very dark in color. We saw one person peering down at us from the top of the mountain. The mountains seemed to accelerate the wind but protected us from swell caused by heavy seas. This provided the perfect place to make our gearbox repairs.

A local fishing boat came alongside; I requested our chief negotiator Johnny "O" negotiate in our behalf. Saying only a few words, our guests left promptly. When asked, Johnny responded these men were Muslim and asking for liquor. Appealing to their religious fervor, they left ashamed.

Johnny machined our transmission output shaft hub using a hand grinder, made keys from allen wrenches, and had the engine working within 24 hours of our arrival. He felt confident we could rely on the gearbox to El Ghouna, some 900 nautical miles away. We decided to remain in the anchorage for an additional 48 hours until conditions improve. I called Ivor, requesting he order a new hub for us that would be sent to the UK and hand delivered to us by the Rally advance team in Egypt. I was asked what was needed most to circumnavigate. The answer is trusted friends we can rely upon.

After the third day we left the Hamish Islands with the Sailing Yacht *Paroo*. Over the next several days, the winds decreased steadily, but we were able to sail conserving diesel for the first three days. We were grateful for Johnny's repairs that enabled us to leave the anchorage safely. Another Rally Yacht forward of our position was less fortunate, losing their engine and the ability to charge batteries. It was agreed the first yacht to reach them would tow them back across the shipping lanes to wait for favorable air. Another yacht was bringing welding equipment to address the source of their problem.

We received sad news that our friends Paul and Liz of *SV Gee Wiz* were dismasted, the Insurance Company would ship their yacht back to England for repairs. Another yacht *Barbas* retired from the Rally due to mechanical failures. When boats are pushed at the speed and schedule that's maintained in a Rally, it's important to have support both on sea and land.

Last evening, we sat in the cockpit under a full moon with glimmering light on the Red Sea. It is a beautiful sight and one that reminds us why we are here. Emotions at sea sometimes become as confused as the weather. One moment it's tranquil, peaceful and relaxing, and the next there's a boiling foreboding tempest that must be taken seriously. We take satisfaction in knowing once we reach Egypt the remaining trip should be enjoyable and relaxing without imposed schedules. The Rally officially ends in Crete early April, and then it's on to Turkey to leave *Tahlequah* for six weeks while we return to the U.S. to visit family and friends.

We received a distress call from another Rally Yacht; the captain was seriously ill but refused a Mayday suggestion. Instead, he elected to remain on board until another yacht arrived to provide medical assistance. We were monitoring his condition by Single Side Band Radio and hoped all would be well. Sometimes making the decision to abandon a yacht is an impossible one to make, regardless of circumstances. In his circumstances, I might have done the same.

Our Crew Johnny "O" and Serun caught several fish including barracuda and prepared them for us immediately. One day we had fish for breakfast, lunch, and dinner, all with Thai spice and rice. The sad part of arriving in Egypt is that Johnny "O" and Serun had to return to Thailand to resume their normal lives and rejoin their families. In addition to their knowledge and professional qualifications, we had a wonderful and fun time having them aboard *Tahlequah*.

We heard a Mayday distress call over VHF radio issued by another Rally Yacht, *Elise*. She was 120 nautical miles north of our position, so we couldn't respond. While at anchor, Gito's son took the inflatable to a nearby reef to snorkel. During this period, the wind and seas increased. Unfortunately, his outboard failed, and rowing for several hours proved impossible. Exhausted, he gave up and was quickly swept out to sea aboard the inflatable. Gito couldn't respond due to reefs between himself and the dinghy, and as darkness was hours away, he wisely issued a Mayday.

The Rally home office was notified; they issued the Mayday to the strategic alliance including coastal patrols. Egypt and Sudan issued an all ships Mayday, and Rally yachts continued repeating the Mayday to all local maritime vessels throughout the evening. Gito decided he would have to find his son alone. He studied local tidal coastal charts during the time of the incident; he began his search the following morning at first light. He spotted the inflatable on a remote island and discovered his son a few hours later. He had taken refuge in an abandoned fisherman's hut.

During this same period, another yacht *Aragorn* arrived in Egypt and immediately set about to charter a private plane to search for Gito's son; fortunately, this was unnecessary due to the happy ending to an otherwise potentially tragic story. If this wasn't enough, there was a second Mayday issued by another Rally Yacht when it began taking on water and was in danger of sinking. Fortunately, two dive boats responded to the Mayday, and the yacht was saved when temporary repairs were made and water was pumped out. Although delayed several days, he did re-join other yachts in Egypt.

Our friends from *Paroo*, David and Sue continued to sail with us for the purpose of providing support if our temporary gearbox repairs failed. Fortunately, our repairs held. One afternoon, I looked over and saw Sue at the top of the mast working aloft in the dreaded boson's chair.

Not bad for a 60-year-old woman, we thought, but then Aussies are a tough breed. Our friends had a habit of wearing their Aussie outback hats when going to town; not to be outdone I decided I'd wear mine as well.

People hollered out, *Hey, you from Texas?*

I'd answer, *No! Australia.*

After 10 days at sea, we slowed down the last night to delay our arrival until daybreak. At sunrise, the Egyptian coastline became visible, and we saw beautiful white sand dune cliffs flowing down to the water's edge. It was exactly as we remembered from the movie *Cleopatra* and expected to be greeted by Mark Anthony rowing to meet us in the Royal Canoe. Finally! We saw a country that exacted the Hollywood version of our expectations. Because we had three weeks in Egypt, we would see the Valley of the Kings, the Pyramids, and fly to Jordan to visit the lost city of Petra and other highlights.

Today, we received an e-mail from the wife of Chris Elmes in Venezuela. Chris was the friend who cared for our condo in Margarita and the family we spent Christmas Eve with. We were saddened to learn Chris was hospitalized two weeks earlier and died that week. His wife had given birth to their son a few weeks earlier. No words can express our sadness or sympathy for his wonderful wife and family.

CHAPTER 20

We arrived in Hurghada, Egypt, at daybreak and tied alongside the Customs and Immigration dock located in a military zone. Our agent hired by the Rally walked us through the many written formalities required of visiting yachts. Next, we were escorted to a nearby Customs and Immigration office where we were queried by the officer in charge. Two yachts before us were asked from which country they came.

"Australia," they responded.

The officer responded, "Six weeks' visa."

Our turn. "America," we responded.

The officer smiled. "Our sister country! Six months' visa for you!"

Touring town, it was common to ask where we came from. Responding *America* always brought a hardy welcome, smile, and handshake. Soldiers guarding the port were friendly, and routinely escorted us to *Tahlequah*.

Two days later, it was time to sail 10 miles north to Abu Tig Marina, in El Gouna. It's a beautiful upscale resort having 35 restaurants, many shops of various kinds, bakeries, hotels, beach, pools, and a health spa. It's the Disney world of marinas, unlike anything we've seen before. Eating out is inexpensive. It's a wonderful feeling to arrive in a new port and be greeted by close friends. We returned to our social mode of having happy hour aboard *Tahlequah* and visiting other yachts for the same.

We found it necessary to separate the beautiful antiquities in Egypt from our daily experiences with current-day shopkeepers. Through worldwide travels, we found it helpful to separate western standards from the values and standards of other lands and peoples, especially true in Egypt. Back to shopkeepers, rarely do we find prices on items in grocery stores and frequently pay a different price at the register each time we return. We're continually short changed in restaurants, shopping marts, and clothing stores. Demanding change can take longer than being served a meal. I purchased a bottle of wine for 45 Egyptian Pounds and didn't get change after giving him 100 E Pounds. Frustrated, I demanded to see the manager; I slowly got my change one coin at a time only after insisting I was still short.

Helen met a young Egyptian shopkeeper who invited her back to his shop for tea. A custom we've grown used to and accept as part of doing business. One evening, we returned to his shop to smoke a water pipe, not an unpleasant experience. His interests seemed to be more in Helen than anything else. Without notice, he would show up at the boat when I was away. At first, Helen was friendly but grew suspicious as this practice continued. When he showed up, she would call her friend on the cell to come over. One day, the shopkeeper asked if I minded if he invited Helen to lunch. That was the last time our Egyptian shopkeeper bothered us.

Helen selected a pair of pants and top in a local dress shop but discovered at the check out she was being charged 40% more than marked. The shopkeeper insisted this was correct. Yes, we walked out. Because our Ship's Papers (Federal Documentation) had expired 48 hours before our arrival our Ship's Agent suggested I falsify the papers to obtain port clearance. A friend managed to obtain my papers legally and forward them to me to meet legal requirements.

When requesting Port Clearance, our Ship's Agent assumed I'd taken his suggestion to falsify papers. He reminded me these weren't legal and demanded I pay him money, so I refused. This didn't discourage him from attempting to overcharge me for services provided, which I also refused. In the end, he patted me on the back and asked I give Helen his best wishes as if nothing had transpired between us. Although this sounds unethical by western standards, it appears acceptable in Egypt.

We heard the word *baksheesh* before arriving in Egypt. Before leaving Egypt, it was a word we grew well accustomed too. *Baksheesh* is a form of alms for the poor, and many feel entitled to this as a gratuity in addition to charges for services rendered. Pilot boats escorting vessels through the Suez Canal demand baksheesh in the form of cigarettes and money; cab drivers request it in addition to cab fare. It's difficult to remember baksheesh is a tradition going back many centuries as a means of helping those less fortunate. By western standards, it's too easy to dismiss this practice as an annoyance.

Johnny "O" and Serun left us in Egypt to return to their homes in Thailand. We gave them a going-away party the evening before, and other Rally boats joined to wish them farewell. Before leaving, Johnny and Serun made a special effort to put *Tahlequah* in good working order. We will miss their company and wish them the best in life.

The adage, you get what you pay for, especially applies to professional crew. We've had several crew persons since leaving the US (volunteer and professional). Of these, half were outstanding, two were of little value, and one had health issues. I've seen yachts dismiss crew at every port; in one case, this happened at sea when a crewmember requested removal by another yacht.

Captain, mate, and crew must be knowledgeable and compatible. In my opinion, compatibility is the more important of the two. When hiring Johnny "O" and Serun, I fully checked out their ship's papers and references. I chose them over equally qualified candidates based on a feeling of compatibility.

Using this process to select crew from the beginning, I would have saved myself heartache, aggravation, and avoided damage to gear.

We had breakfast each morning at a French bakery in the marina and frequently went into Hurghada to purchase supplies and souvenirs. Sand storms are new to us, but not Egypt. We had so much sand aboard *Tahlequah*, I considered purchasing a child's beach bucket and shovel to play in the cockpit. I periodically wash the boat down to clear the sand, but it finds its way into the cabin, sheets, and sails, and even the winches must be covered with plastic bags to prevent damage.

On March 13, we celebrated Helen's birthday by gathering at an Indian restaurant with other Rally boats. A French bakery produced a birthday cake resembling a giant French tart, and it was good if not familiar. We ate and drank beyond usual requirements, but it was Helen's birthday. It was one year ago we celebrated Helen's birthday in Galapagos; how time flies and life moves on.

It was time to get a haircut. The last several haircuts had seen my hair length reduced several inches; so much for the long hair I promised myself after retirement. Realizing it was a losing battle and knowing Helen wanted me to look like everyone else, I finally agreed. Eyes shut, teeth gritted, an Egyptian barber cut off my ponytail. As my hair fell to the floor, I heard three French women saying, *uhhhhhhhhh!* I knew it was final.

There's an interesting Egyptian technique of removing unwanted facial hairs with dental floss, but don't ask me to explain this one. Afterwards the barber attempted to persuade me to look into a mirror, an invitation I flatly refused.

Helen, hope you're happy now! Everyone says I look younger, so how come I feel neutered?

March 14, we flew to Jordan with our good friends Dick and Leslie York of the *SV Aragorn*. It was a four-day trip of unparalleled highlights, put together by the master of organization, Leslie. The excursion was escorted, catered to individual needs, and included airport and hotel transfers, daily private tours, and meals. Although Jordan is a country of only five million people, there's a 91% literacy rate, many speak English and consider themselves international citizens. The Bedouins continue to inhabit the desert as nomads, living in tents and moving freely through the country.

The first day in Jordan, we traveled to Mount Nebo, where Moses spoke the word God and spent his final days. There's a monastery atop the mountain (several hundred years old), which has beautiful mosaic tiled floors. A large group of Jewish businessmen were being guided through the monastery.

I asked our guide if this was a usual occurrence, and he responded, "Of course, these people are our neighbors and friends." His simple, quick, and honest response summed up what was to become my feelings for the Jordanian people. Traveling the world has given us a different perspective of people and places than the TV news programs.

This evening, we received a special treat when we checked into the Taybet Zaman Hotel, a converted Byzantine Village. It's a five-star hotel complete with stone houses and winding cobblestone streets restored to previous glory. The food was exquisite, served in a Byzantine dining hall. The following morning, we awoke to discover we were perched on the side of a mountain overlooking the hidden city of Petra, one of the wonders of the world. My hip was increasingly becoming a problem. Refusing to use a cane, I brought a smart-looking hiking stick to help walk.

The city of Petra was immortalized in the film, *Indiana Jones and the Temple of Doom*. I wore my Indiana Jones hat (really an Australian Outback hat) as I rode horseback into the city with Helen. The city of Petra was founded in the first century BC and reached its highest population of 3,000 inhabitants. The city was founded and inhabited by the Nabataeans and supported the major trade route. The city was well hidden and protected by its natural surroundings, and it was impenetrable for two centuries and was forgotten until modern times.

Even by modern standards, it's an advanced city, having underground clay water pipes and viaducts carved into the descending valley floor. The city is divided into two sections, the Valley of the Dead and the Valley of the Living. The Valley of the Dead contained colossal tombs carved into the mountainside intended to entomb royalty and important persons of the day. The Valley of the Living consisted of homes carved in the limestone cliffs and stone houses erected along the canyon floor. Fifteen percent of the area has been excavated, but much work continues by teams of American and UK archeologists.

In 1993 the Jordanian government offered to relocate the last inhabitants of Petra to new housing, offering them tourist employment opportunities. Petra was restored to its former glory and is being rediscovered by people from around the world.

There are many colossal tombs and temples carved into the valley walls, each being more magnificent than the next. The creators of these tombs began carving from the top of the mountain, not the bottom as might be expected. Large columns supported immense structures whose facades were decorated with statues of the gods. Although the tombs were robbed in early antiquity, the architecture remains a tribute to its creators. Petra is a city unlike anywhere else and should be visited by those wanting to see another wonder of the world.

Inside the City of Petra, we talked to a Bedouin (nomadic desert people), who invited us into a cave for tea. The cave had hand-woven carpets on ceiling and floor with camel saddles and cushions to sit on. Most Jordanians speak some English; we were able to discuss world issues and the countries we came from. We talked about King Hussein and how we regarded him as a diplomat and statesman. Our Bedouin friend shared that although he was a businessman, he needed to spend time in the desert; it was his heritage and tradition to do so.

The city of Petra had many donkeys, horse-drawn carts, and horses and camels for transportation. Young children collected stones and broken fragments, attempting to sell them to tourists. Not wishing to encourage the destruction of antiquities we avoided these situations and instead turned our attention to purchasing legitimate merchandise.

That evening, we returned to the hotel and enjoyed a new experience, a genuine Turkish bath in our Byzantine Hotel Village with our companions and friends, Leslie and Dick. We were informed everything needed would be provided, so we didn't bring bathing suits. Entering the Turkish bath was an experience of going back in time. A man provided each of us sheets with instructions to put our clothes into a special locker provided for this purpose. We were escorted into a large marble room, complete with a raised heated marble table. Off this room were several steam rooms also of marble. The steam was overwhelming, but our bodies adjusted quickly.

Following steam treatment, we returned to the center room, and we lay on the large heated marble table and gazed up into the domed ceiling above us. Later, a man entered and instructed me alone to follow him. I lay on the floor while he sat next to me, the towel disappeared, and he began exfoliating my skin with a special sponge. Next I was given a massage that relaxed every muscle in my body. When finished, he requested I send Helen alone. Helen, Leslie, and Dick received the same treatment.

Leslie described how she surrendered her sheet saying, "Well, if that's what you're supposed to do."

Dick returned and said, "Is he supposed to rub everything?"

Later we were taken to a Jordanian restaurant where we prepared our own dinner. A chef assigned each of us a task in preparing one part of the meal. We made chicken cooked in yogurt, lentil soup, salad (my task), hummus, and many other special treats I can't spell or pronounce. The end result was a gourmet dinner, with the help of our Jordanian chef and others; it was a culinary delight. I admit the salad is the first thing I've ever prepared.

Drinking wine by the glass, we didn't realize Jordanians don't drink wine; the bottles we didn't finish were later discarded and added to our bill.

Sand storms are a common occurrence in Jordan. Dick and I purchased Coiffeurs (popular mid-eastern head dress) to protect us from the blowing sand and to keep us warm during the cold mornings and evenings while trekking. Needless to mention, Dick and I look Irish-German, but no one paid notice to our unusual appearance as westerners. It's now easy to understand how Lawrence of Arabia fit in. In time, we forgot we were wearing coiffeurs because of total acceptance.

The following day, we drove to the Wadi (in a four-wheel drive), a desert-like area where *Lawrence of Arabia* was filmed and more recently *The Red Planet*. The sand had a reddish tint and was an arid landscape with beautiful surrounding mountains. We climbed a cliff crevice to see petroglyphs thousands of years old. Returning to the four-wheel drive, we discovered our driver, an elderly Bedouin man, preparing tea for us in a small teapot heated over a fire. It added to the authenticity of the moment, and we were delighted at the experience. In the background, we saw Bedouin tents and camels.

Next, we drove to the Dead Sea and swam using the beach of a local hotel. Once in the water, it was impossible to do anything but float due to the amount of salt, contributing to our buoyancy. Unable to get her feet down, Helen began floating away, an interesting phenomenon that had a happy ending; yes, we saved her again.

Getting water in your mouth or eyes can be compared to the experience of an acid burn. Signs everywhere warn unsuspecting tourists to the hazards of the Dead Sea. We dug black mud from the bottom and smeared it on our bodies, a common tourist practice of those wishing to look younger.

Unfortunately, we waited too long to wash the mud off, and yes, some of us cracked. Walking to the beach showers in slow motion, we frantically washed the mud from our bodies, but we got off what was coming off and dressed to leave. We convinced one another we had erased 20 years from our appearance and were now wrinkle free. From our vantage point on the Red Sea, we could see Jericho, Bethlehem, Jerusalem, and the Palestinian area from the beach, a moving experience. Our guide described how both Jordan and Israel cooperated together in patrolling this area of the Red Sea.

On March 17, we returned to Cairo, Egypt, for one full day en route back to the Abu Tig Marina. In Cairo, we toured the Pyramids, Sphinx, and Cairo Museum with an Egyptologist Guide. Although the City of Cairo is encroaching upon the Pyramids, you get the feeling you're in the dessert. The generators providing electricity inside the Pyramids weren't working. We persuaded a guide to give us a private tour using his torch. Once inside, the man disappeared holding Helen's hand. Left to my own resources, I felt my way through the narrow and shallow passageway until as luck would have it, the lights suddenly illuminated the passageway. Hopping along, I managed to catch up to my fearless guide and Helen still holding hands.

Reaching the main burial chamber located in the center of the pyramid, we saw the sarcophagus. Our guide invited me to lie inside but not knowing his exact intentions, I declined the offer. It was a memorable moment, standing inside one of the wonders of the world. Unlike many civilizations, Egypt was preoccupied with the next life, having little interest in the present one. Our guide asked if I was happy with the tour and without waiting for a response, he requested I pay him.

I politely responded, "Not until you successfully get us out again." Next we traveled to the nearby Sphinx and buildings associated with the rites of mummification. We were informed recently discovered chambers exist inside the Sphinx yet to be opened and are now subject to x-ray studies.

A major highlight of our trip to Cairo was the Cairo Museum. Our Egyptologist Guide explained every major detail of this incredible museum. The King Tut exhibit alone made the visit to the Cairo Museum worthwhile. On exhibit were the young Pharaohs (he was 19 years old when he died) three burial boxes, each with gold gilding, three caskets of breath-taking beauty made of solid gold to the likeness of the boy Pharaoh himself. Also exhibited were the Pharaoh's gold mask, linens, necklaces, belts, finger and toe covers, and jewelry, all in perfect or near perfect condition. The entire tomb discovered by Carter was undisturbed when discovered in 1922. We saw many life-like dolls and statues of important persons, priests, scribes, and architects of the time.

On two separate occasions, we joined a Rally tour to the ancient city of Luxor where the Valley of the Kings and Karnack Temple is located. This area is referred to as the New Kingdom versus the Old Kingdom (Pyramids). A bus caravan (50 buses) left Hurghada under tight military escort. The major road to Luxor was closed to pedestrian traffic with armed military patrols stationed at all crossroads. Although the threat to foreigners is considered a minor one in Egypt, this show of force appears to be a form of reassurance to tourists. At no point during the three-hour drive did the caravan stop until we reached our final destination, the City of Luxor.

In Luxor, we visited the Karnak Temples that took 2,000 years to construct and until recently was buried beneath 21 feet of mud resulting from flooding of the Nile. It's a spectacular setting with many large columns, statues, and high needle-like monuments called obelisks. Prior to a great earthquake, the roofs continued to be supported by the columns; although today, few survive. Hieroglyphics adorned the columns and walls everywhere. Each succeeding Pharaoh added to the temple during his reign.

Our Egyptologist guide was extremely knowledgeable. Even more importantly, he was proud of his heritage. His descriptions of Egyptians and their history inspired us as he talked about their accomplishments in science, engineering, math, and the culture of mankind.

We visited a Papyrus Institute where paper is made from papyrus in the traditional manner and adorned with colorful reproductions of scenes copied from the temples. We purchased many of these for family gifts, one to be on exhibit in *Tahlequah*'s main cabin. We watched papyrus plants being cut into flat stalks, soaked in water, pressed into cross-matted patterns, and finally dried under pressure resulting in a strong indestructible paper of the highest quality. The color of the paper can be controlled by the length of time soaking in the water.

Next stop was Valley of the Kings where hundreds of motor coaches converged into one massive parking lot. The valley itself was a barren dessert void of vegetation. The soil appeared so loose, I feared it would collapse upon us. We entered three tombs hidden beneath the valley floor. Although these tombs were likely looted by the very people who participated in the burial, the inscriptions and paintings bore evidence of their beauty and significance. We entered the tombs passing down steep staircases to the wide passages below. The main passage always ended at the burial chamber still containing the sarcophagus. It was a remarkable feeling passing through these passages examining the frescoes above and to the side of us. The elaborateness of each tomb depended upon the Pharaohs years in power and took easily 20 years to complete each one.

Later we took a Felucca (traditional Egyptian sailboat) ride along the Nile. The Felucca has a Gaff-like rig, wooden mast, and steel hull. We saw fishermen using nets, many children paddling and waving to us, and farmers working in the fields along the Nile. The following morning, we took a horse-drawn cart ride through old Luxor, seeing narrow streets filled with vegetable carts and street vendors of all kinds. Mule carts were common in the streets selling vegetables and fruits. The streets were so narrow, our horse-drawn cart barely fit through the bustling streets.

We visited a spice shop, accepting a cup of tea from the shopkeeper who explained each spice and herb in lengthy detail. Fearful we were about to be duped into purchasing expensive spices, we attempted to leave the shop. The master shopkeeper managed to persuade the women to purchase small quantities of simple spices with reassurance they were not expensive. When time came to settle our bill, we discovered the cost for our inexpensive spices was more than expensive. We walked out in mass only to be pursued down the street to our horse-drawn cart. After much yelling and screaming, we fled the alley to the dismay of our shopkeeper. We informed our driver of our dissatisfaction in his choice of shopkeepers, but in the end we tipped him well.

We returned to the Karnak temple illuminated at night to emphasize its beauty. Seeing the ruins at night is more spectacular than seeing them during daylight. Again, Adel's (our Egyptian guide) enthusiasm in describing historical events was captivating and had us enthralled the entire evening. At the entrance to the Karnak temple, there was a light show. There was music and dialog inviting us to enter through the columns. As we walked through each area, it was illuminated with a dramatic explanation of where we were and its history.

In the evening, we returned to the hotel. As we exited the elevator, a man standing nearby rushed to our room and entered, alerting four other men inside our room. I waited in the hallway demanding an explanation as they exited our room.

Hearing, "No English, No English," I called the hotel manager demanding an explanation and requested a police report be filed.

The manager came to apologize and offered several flimsy explanations that sounded doubtful. He asked I not request the police; he would be fired for allowing travel agents to inspect my room believing the room was unoccupied. We considered leaving the hotel, but due to the hour, it was impractical. I made several phone calls complaining about the incident and barricaded the doors until daylight. In hindsight, since Egypt is a police state, they may randomly search hotel rooms and look for who knows what, or then it could just be travel agents inspecting rooms at midnight ... hmm!

CHAPTER 21

We were delayed leaving Abu Tig Marina several days due to high winds and sand storms. Knowing the end of the Rally was near, skippers became more cautious concerning weather and departure schedules. Several yachts sustained damage causing the rest of us to be more conservative. A weather window appeared and the Rally left for a two-day sail to the Suez Yacht Club at the entrance of the Canal. Following two days at the Yacht Club, we were ready to resume our journey. Our Ship's Agent, Prince of the Red Sea, wore a white cape and took responsibility for our social events. A special dinner was organized at a local hotel where we partied for the evening. At the end of the evening, I had difficulty standing because I'd forgotten my cane. Patrons sitting at a nearby table took great amusement in my difficulty, believing I was drunk.

One morning we received a radio call informing us a sand storm would soon be blowing across our anchorage at the entrance to the Suez Canal. In the distance, we saw a haze slowly making its way towards us. Minutes later, our Ship's Agent rowed towards us requesting we double our mooring lines and cover our winches with plastic bags. Ten minutes later, we were belted with 40-knot winds and blinding sand. Although tightly secured, we strained the moorings, coming within 10 feet of the vessel tied next to us. The *SV Paroo* behind us broke their mooring lines and began swinging wildly. Within a short period, the sand storm had blown past.

The Suez Canal marks another major milestone in our circumnavigation. Unlike the Panama Canal, the Suez Canal has no locks linking the Mediterranean and Red Seas, so there's no change in level. Work began on the Canal in 1859 near Port Said and was completed in 1867 after being suspended several times due to engineering and political issues. Unfortunately, due to the 1967 Israeli conflict, the Canal was closed for several years and not reopened until June 5, 1975. Since that time, the Canal has been enlarged and updated. Canal traffic never regained its pre-war traffic levels.

We were officially measured; papers cleared and prepared to leave in two groups over a period of two days. After the first group of sailboats departed, we were informed we must leave within 50 minutes or be delayed another five days due to a fleet of US warships coming through the Canal. We were informed this was privileged information and should not be shared. Within minutes, a non-Rally boat over VHF radio demanded to know why Rally boats were privileged and others were not. Prince of the Red Sea responded this was a security issue and could not be discussed on VHF radio.

Due to a shortage of pilots, the Canal Authority provided one pilot and tug to escort the flotilla through the first half of the canal. Motoring at seven knots, we reached the Yacht Club located midway through the Canal at sunset. The following morning, we were off to complete the Suez Canal transit and enter the Med for a three-day trip to Crete, Greece.

The Canal operates allowing northbound traffic to transit morning and southbound traffic the remaining time. By hugging the edge of the canal, we avoided all canal traffic and maintained good speed throughout the day. The transit is straightforward and without delays except for the overnight stopover in the center lake.

I didn't follow the boat ahead as directed due to its slow speed; it would have resulted in long delays. Instead, I followed a faster Rally boat and radioed to inform the pilot aboard. At the end of our transit, we passed alongside the pilot boat, and Helen threw a carton of Marlboro cigarettes with $30 US attached, his baksheesh. The pilot smiled, waved, and appeared grateful, and everyone was happy. Because we never saw US warships in the Canal, we theorized we were misled for the purpose of making room for yachts behind us.

Leaving the Suez Canal, we set a course directly for Aghios Nikolaos Harbor, Crete, Greece. During the three-day trip the winds averaged 15 knots, getting as high as 35 knots. Knowing this was the official end of the Blue Water Rally, we had mixed feelings. A part of us was pleased we wouldn't have to rush from country to country; another part would miss the camaraderie.

On the morning of April 8, we saw the beautiful island of Crete in the distance; snow was visible at the highest elevations. The town of Aghios Nikolaos is typically Greek, with white buildings immaculately maintained rising above the waterfront to the hills above the harbor. Outside the harbor entrance, we were met by a marina dinghy sent to direct us to our berth. Rally boats were there to greet us at the dock; it was a reunion of old friends.

After usual formalities, the Rally brought us a bottle of champagne and oranges. After others arrived, we gathered on the dock to drink champagne and celebrate our great accomplishment. For those who began the Rally in Gibraltar, this represented their return to the European Community. *Tahlequah* has to complete the Atlantic crossing, and we look forward to completing it one day.

The next few days, the Rally organized several bus tours bringing us to old Turkish ruins, Byzantine forts, beautiful fishing villages, and towns specializing in hand-sewn lace. On the second day, we were invited to a dinner party given by the Mayor of Crete at the Hermes Hotel. For many, it was the first time we'd worn jackets and ties in a few years, and some were hard to recognize.

On one trip, we took a boat ride to the island of Spinalonga, an old Venetian fort and later leper colony. Seeing the leper cemetery causes one to stop and think about the many people once living here. Within the fort were medieval winding streets and original storefronts and stone houses. It's a wonderful place to spend a sunny afternoon walking around the island.

We shared a car with *SV Paroo* to visit the Herakleion Museum. On display were the pottery and frescos discovered at the Minion ruins of Knosos, and they were fabulous. After the museum, we were fortunate to visit the city of Knosos, itself a remarkable and unusual archeological site.

Controversy exists because the excavator, Dr. Evans in the early 1900s purchased the property, excavated the site, and reconstructed many of the buildings using concrete. This was done during a 35-year period after which he donated the entire site to the Greek people. From a layman's point of view, it gave us a remarkable opportunity to see more than rubble and actualize life here thousands of years ago.

At the end of the day, we drove to the highest peaks of Crete. There were many Greek villages unchanged by time. Sheep herding, farming, glass blowing and pottery remain here a part of normal life. We see old ruins and Greek Orthodox churches everywhere. Concerned that the roads were narrow, dirt, and had no guardrails even at these elevations, we returned to sea level before dark.

The time came to have our final Rally celebration. We were taken to a restaurant in the countryside owned by two Greek brothers who loved music, Greek dancing, and spirits. Through a lottery, each yacht drew another yacht to say something about; some were funny, some were warm, and all were memorable. The following was a poem presented by Midge of the *SV Regardless*. Midge helped stabilize Helen en route to the Phuket International Hospital following the tsunami. During the poem she began to weep but found strength to continue.

Tahlequah
There's a cruising yacht named Tahlequah,
and beautifully fashioned too.

Almost all around the world she's been,
with Ed and Helen as crew.

Sometimes contrary and unwilling to budge,
she's made it to Crete without too much of a dredge.

Along the way she's served as home for those she's had aboard,
Muesch generosity to others always very Broad.

Traveling with other yachts across the Pacific Blue,
the Temors and the Java Sea and uppity Indian too.

And then on Xmas Day plus one there came a tidal wave,
the havoc all around us sent many to their grave.

But Tahlequah with rope on prop made it through the wave,
with Mike aboard and Tom York there to help him get it off.

Ed and Helen on shore were nearly washed away,
miraculously surviving to tell a tale of horror and disarray.

It was our good fortune to smooth their path to hospital along with many others,
a great relief to all their mates to know they would recover.

A damaged back, a gimpy leg, a walking stick at his side,
Ed wobbles along at a steady gate with Helen as his guide.

Their mild and friendly manner, a joy to all they meet,
may they travel safely onwards in the green
boat of no fleet, Tahlequah.

By Midge of the SV Regardless

We sat at a table with Ivor, Bernice, and Franco (*Safari*) discussing old times and future plans together. Watching Demetrius, a young Greek owner of the restaurant, dancing to the Greek music inspired us all. Demetrius gave us an unbelievable demonstration of Greek dancing; he picked a dining table up with his teeth and danced around the room to Greek music. Ouch! I'd have left my teeth on the table if I tried that. He performed other feats; the table lifting did it for me.

One more drink of homemade Raki, and Ed and Helen were dancing with everyone else. Led by Demetrius, the Rally danced together into the night, some performing their own feats of wonder.

I'd arranged a dinner meeting with a Blue Water Rally Director to discuss my acquiring the Rally as an investment. I was informed, due to complications within the Rally at this time, this couldn't be considered. Disappointed, the subject was dropped.

We shall always remember the yachtsmen and yachtswomen of the Blue Water Rally Fleet. The Brits with their hardy spirit, always willing to endure and never surrender, are a tough lot indeed. Our South African friends were always there for us and we'll always be close to them. I think of our Swiss friends, especially Margit who liked to dance with Helen, telling her in broken English, *I'll be the man*, and our Dutch, Spanish, Australian, New Zealand, and American friends, who'll always be with us in spirit.

Each and every one of these people helped us in so many ways, it's impossible to pay tribute to each of them. Sometimes towing, sometimes repairing computers, sometimes mechanical, sometimes lifesaving, boat saving and sometimes just a pat on the back when we needed one the most: May each of these wonderful people continue finding new challenges, adventures and dreams to follow, may they always be happy. Without you, we would never have been able to realize our own dreams. We owe each and every one of you so much!

The day came for the Rally to officially end, and each of us would now move on with our lives and go our separate ways. Some of us planned to return home, some to put their yachts up for sale, and others would begin their next adventure aboard. After tearful goodbyes, we started our engine and motored out of the harbor. Raising our sails, we began our journey to Bodrum, Turkey, and our next port of call and new adventure.

CHAPTER 22

We arrived at Bodrum Marina in the morning and were escorted to our berth by a marina dinghy. The day of arrival, our good friends Dick and Leslie invited us for drinks and dinner knowing we were tired from an all-night passage.

The City of Bodrum is very Bohemian. It was once a place to send dissident authors, poets, and artists outside the political system of the times. Today, Bodrum is a major tourist area and is a wonderful place to wander the streets shopping, eating at street cafes, frequenting shops of all kinds, and there is a Seamen's Club open to all. There are hundreds of beautiful Gullets, very large wooden sailboats with brilliant bright work, lining the waterfront awaiting day guests. These boats are magnificent, and *we were* hoping to have the experience of sailing on one before leaving Bodrum. St. Peters Castle is along the waterfront and houses a museum *we were* planning to visit while here.

We purchased airline tickets for our long-awaited return to the US to see family and friends. Our major goal was to complete all postponed repairs to *Tahlequah* deferred due to Rally schedule. We lined up mechanics, electricians, and sail makers to complete work during our absence. We've purchased a new dinghy and outboard in preparation for our forthcoming cruising in the Med.

Turkey is unlike anywhere else we'd been to date. Turkey is a Muslim country and may become the first non-European country to become an EU member within ten years. Turkey is where the earliest Christian towns were established. Because of many faiths throughout the centuries, there's a tolerance of other religions in Turkey today. Religion and government are separated constitutionally; the country is governed by Parliament on democratic principles. The economy is healthy.

Many Europeans have chosen to relocate here, especially British, Germans, and French. The Turkish people are customer oriented and wish to please. We never ask more than once for information or assistance; it's provided immediately. Turkey is a proud country; Turkish flags fly everywhere.

As we arrived in Bodrum Marina, the staff requested we remove our Turkish Courtesy Flag, because it was incorrect. It was explained the moon and stars were not correctly positioned, the flag must be replaced. Surprised, but unwilling to make an issue, we replaced the flag. Consumer costs have doubled in three years, and although it remains inexpensive by EU standards, it's expensive by American standards.

We returned to the US in April for six weeks to visit family and friends. En route to the US, we stayed one extra day in Istanbul to tour the city. The city of Istanbul has a rich culture and history. We walked winding streets with endless stonewall fortifications. We visited the famous Blue Mosque, the largest mosque in the world; it is 379 years old. Leaving shoes outside, we walked through the immense structure. There are several balconies with hundreds of stained glass windows overlooking the prayer area. The floor has thousands of small Turkish prayer rugs, where worshipers prostrate themselves kneeling toward Mecca to pray several times daily. There's no statuary or adornments to distract worshipers from prayer, and it's simple but beautiful. The second balcony has blue mosaics, hence the name Blue Mosque. There are several large domes supported by immense columns. Large candelabras now electrified are suspended by cables from above. Outside the mosque are several areas for washing and purification, a common cleansing ritual before prayer. Beautiful gardens surround the outside of the mosque.

A bazaar-like atmosphere on the streets exists outside the Blue Mosque. Shopkeepers invite tourists passing by into their shops for Turkish coffee or tea. We've found, unlike many countries, shopkeepers don't apply undo pressure here. It's acceptable to enter a shop, have tea with the proprietor, and buy nothing. Well-dressed men wearing suits walk outside and inside the mosque. These men, after making clear they're not guides, will answer any questions you have about the mosque or Muslim religion and spend an entire afternoon answering your questions and satisfying your interests.

One elderly and knowledgeable gentleman walked with us for much of the afternoon showing us everything there was to see both inside and outside the mosque. At the end, he asked if we would have an interest in seeing his carpet shop, and of course, we accepted. Once inside, we were invited to sit on a stack of carpets and enjoy wonderful Turkish tea while several members of his family laid out one carpet after another on the floor in front of us, each hand-woven carpet was more beautiful than the next. Each carpet design had an accompanying story. Having these stories revealed to us for the first time proved fascinating. After explaining that although we intended to buy one or two carpets for our cabin sole, we didn't have the dimensions with us and couldn't purchase anything today. His disappointment never showed, and he insisted we have more tea and tell him about our world adventures. Every human encounter experience we've had in Turkey has been the same.

We transported many boat parts on our return trip to Turkey. We were pleased that *Tahlequah*'s projects were completed, including sail repairs, cushions, zippers, gear box, alternator, new dinghy/outboard and finally a coat of varnish. The outstanding projects including prop shaft replacement, new motor mounts, and turbocharger will be completed before August.

We contracted to relocate from Bodrum Marina to D-Marin in the town of Turgutries, 15 miles further north along the coast. This marina has a swimming pool, yacht club, haul out facilities, restaurants, a theater, and is conveniently located within town. We signed a one-year contract as *Tahlequah* will be on the hard for a seven-month period during our absence.

Before leaving the city of Bodrum, we toured St. Peters Castle, a beautiful and well-preserved castle built by Spanish Crusaders and later occupied by British knights during the crusades. Unfortunately, the stone blocks used to construct the castle were taken from one of the seven wonders of the ancient world, the Mausoleum. All that remains of the mausoleum today is its foundation. Today the castle is in excellent condition and remains entirely intact. Inside the castle is housed a beautiful exhibit of antiquities recovered from recently discovered wrecks off the coast of Turkey. The exhibits include exquisite glass, pottery, tools, jewelry, coins, and one excavation site that has been reproduced as it was first discovered in 130 feet of water. Before leaving the castle, we enjoyed another cup of tea.

Helen purchased Turkish carpets for our cabin sole, contributing to a homelike atmosphere aboard *Tahlequah*. These we purchased from our Turkish friend Tolga living in Bodrum. Tolga and his wife own a travel agency and other businesses in Bodrum. One of these businesses is a Turkish carpet shop located above his travel agency. Although invisible from the street, we expect he sells these carpets retail to others. Helen selected a special prayer rug for our dear friends Ivor and Bernice, who we will visit this August in South Africa.

One benefit of having a marina contract is we enjoy the marina amenities, including yacht club, swimming pool, watching the evening news, and occasional dinners in local restaurants. We've sailed the coast of Turkey visiting many beautiful anchorages and local communities before returning to the marina to renew the process of leisure and relaxation. We've discovered this method of sailing puts the least amount of wear and tear on my hip, and we are able to see the country.

One evening anchored off the town of Yalikavak, a man in his late 60s swam towards our boat, circling it twice. Because we were far from shore, I invited him aboard to rest. He smiled, said something in Turkish, and began his return trip to shore. Moments later, it was apparent he was exhausted; I called again and invited him aboard. This time, he graciously accepted our invitation and returned to board *Tahlequah*. Although he spoke no English, we communicated through gestures. Helen asked if he were Muslim, he responded with a gesture indicating yes, but not a strict Muslim, he insisted. We learned he lived in a house overlooking the sea and swam daily to remain healthy. I offered to return him to shore in our dinghy, but he would have no part of it. After one beer, he dove back into the sea and returned to the beach. We've taken notice Turkish people love the sea and routinely swim far from shore, regardless of age or in some cases condition. Their culture is so entwined with the sea their comfort level seems far beyond what we normally see in other countries.

We took a two-day tour to the ancient cities of Ephesus and Pamukkale. The total cost for both Helen and me, including meals, entrance fees, hotel, and transfers was only $125 per couple, certainly one of the best deals we've experienced. The benefit of a tour is we met wonderful people from the UK, Ireland, Poland, Russia, and Norway. We made several new friends and exchanged e-mail addresses. Having someone explain to us what *we were* seeing is worth the cost. For this reason we avoid car rentals and self guided tours whenever possible.

Touring the ancient city of Ephesus was an extraordinary experience. Ephesus is where St. Paul, St. John, and Mary lived after the death of Christ. The town dates back to two millennium BC. Many ancient Roman cities can be considered modern even by today's standards. The city has marble streets, an amphitheater seating 25,000 persons, a massive library, public baths, and beautiful statuary.

Moving on, we visited Pamukkale, where thermal springs created white calcium cliffs atop a mountain through the centuries. The calcium deposits look like snow. They are hard and rock like. Thermal pools are scattered throughout the cliffs and *we were* permitted to walk along the cliffs barefooted. Behind the cliffs is the magnificent ancient city of Hieracropolis where a quarter of a million people once lived during Roman times. The city was massive, and much remains today including three amphitheaters, public bathhouses, hundreds of homes and shops, libraries, schools, and public toilets. The cemetery is large consisting of hundreds of beautiful tombs adorned with carvings and statuary. Civilization's first discovered street sign remains, directing people to the local Roman brothel.

Inside the city, we swam in Cleopatra's pool, so named because it's rumored she once swam there. The pool is a natural thermal spring with the remains of beautiful broken marble columns scattered on the bottom. Swimming here is a wonderful and refreshing experience. We find ourselves reading books on Greek mythology; the more antiquities we visit the more we become interested in discovering their history.

We sailed along the north coast of turkey stopping in beautiful anchorages. Wooden charter boats called Gullets take tourists on weekly sailing excursions. Gullets are beautifully crafted traditional sailing vessels, reminiscent of the early days of sailing. Gullets continue to be manufactured in Turkey and are exported around the world. One new Gullet was 125 feet long.

There are many sailboats and yacht clubs here with no shortage of sailing competitions. Our Marina Yacht Club is currently hosting two major Regattas this week, and the parties are open to all Yacht Club members and their guests.

Like Greeks, Turks enjoy dancing, often men together. We befriended a young Turkish captain on a private yacht; we routinely socialized on each other's vessels. One evening, he cooked dinner for Helen and me, and we returned the favor in kind. When his girlfriend joined him for a few days, we enjoyed her company as well. I've become used to men kissing me on the cheek. Turkish men and women are very passionate, warm, and friendly. In Turkey, it's not necessary to lock our boats or dinghies when we leave the yacht.

We were saddened to learn of two bombings in Turkey this week with the loss of two Europeans. A Kurdish militant group claimed credit for the violence. I learned this small Kurdish militant group has been active many years in hopes of forming an independent homeland on the borders of Turkey and Iraq. We've encountered no anti-American sentiments in Turkey. We feel as safe here as in any western country—as safe as one can feel these days. Europeans have discovered Turkey, making it a major tourist destination. Hopefully one day Americans will discover this fascinating country.

CHAPTER 23

We remained at D-Marine to celebrate the Fourth of July with other American yachts. *Tahlequah* was formally dressed (signal flags) for the occasion. During the month of July, we explored the north coast accompanied by another boat, the *SV Almedes* and our friends Don and Sue from Southern Cyprus. Although a short trip, we explored several anchorages. At the town of Torbali, we anchored and went ashore for drinks and dinner. The town is a picturesque one where young Turkish couples and families come for holiday. A waitress asked if we were from those boats as she pointed toward *Almedes* and *Tahlequah*, and we answered yes, and she disappeared inside.

Returning, she was accompanied by the owner of the hotel. He explained in his youth he worked aboard boats. He offered us showers and invited us to return for a buffet breakfast in the morning, normally reserved for guests. This has always been our experience with Turkish people, wonderfully hospitable and friendly. After a week cruising the north coast, we returned to Turgutreis. During our absence, another article describing our tsunami survival experience appeared in a local newspaper. Several people came to speak with us about the experience.

Taking the ferry to Kos, we spent an entire day exploring this Greek island. Although Kos is less than a 45-minute ferry ride from Turgutreis, it's typically Greek, differing greatly from Turkey. Greek architecture, although white frequently, has red tiled roofs scattered or clustered along the shoreline. Although our main purpose was to purchase duty-free liquor, we did want to see Kos, the home of Hippocrates. We looked for the spot where Hippocrates is reputed to have written his now-famous oath. We discovered the tree located in front of an abandoned mosque. Large limbs had collapsed and rerooted themselves into the soil, giving an almost grotesque appearance to the tree.

An iron fence protected the tree from vandals. Although not the original tree, it appeared to be very old and in poor condition. Small shops located along the outside of the mosque sold reproductions of the Hippocratic Oath in many languages, each having a leaf from the tree inside the plastic protective covering. We purchased two, one for our grandson and another for the doctor who invited us to give a tsunami presentation in Durban, South Africa.

Although only on the island of Kos for one day, we made the best of our time by taking a tram ride around the island. After returning to the waterfront we set off to return to those places that interested us most. We stopped to have lunch adjacent to an area having early Greek and Roman ruins nearby. After lunch, we toured the ruins watching archeologists unearth artifacts, removing dirt to filter through screens, and we took many pictures of sub-structure levels of columns and building wall foundations. We've grown accustomed to seeing ruins throughout Greece and Turkey and continue to admire the early civilizations that constructed them. Had it not been for the great earthquake that occurred in the seventeen-hundreds many more of these ruins would be standing today.

Because we were given Yacht Club memberships as a condition of our one-year contract with D-Marin, we were invited to the several annual Regatta parties without actually participating in the Regattas. These parties were extravagant even by US yacht club standards. Free food, drinks, music, and dancing were all part of the day's activities and continued throughout the night. We distinguished ourselves by being first at the buffet and drink lines.

A few days later, our good friends David and Sue of the *SV Paroo* arrived in Turgutreis. Because *Tahlequah* was now on the hard for the season, they invited us to sail the southwestern coast of Turkey aboard *Paroo*. For 12 days, we enjoyed both their company and anchoring in many different locations along the coast.

We frequently find ourselves in the company of Gullets, with tourists aboard. These wooden boats are beautiful in appearance but rarely raise sails or have no sails. They tie stern lines ashore when there's no apparent reason to do so. We've grown accustomed to their habits and welcome the company, especially when they play Turkish music for the entertainment of their paying guests.

Following our Pilot Book, we couldn't leave this area without visiting Cleopatra's Beach. Local folklore claims Cleopatra brought sand from Egypt so Mark Anthony would feel at home when bathing here. I admit the sand is unlike any I've seen elsewhere in Turkey. Regardless, the story is a romantic one.

Entering through reefs, we anchored in a bay inside several islands in the company of more Gullets. Because this is a National Park, a small fee is charged to go ashore. Again, there are ancient antiquities that make exploring the island interesting. The unfortunate inhabitants here sided with the losers of a great battle and were later sold into slavery by the Spartans. Because so many tourists visit here aboard day charters, Cleopatra's Beach is very crowded. Helen and Sue swam before returning to the boat. Fortunately, before dark, all day charters return to the mainland leaving the anchorage to us. Each evening, we enjoyed sundowners on deck during sunset.

Many nights are spent in secluded areas with green cascading hills and cliffs descending into the water. Often, we'll anchor off of ancient ruins that invite exploring before leaving. One such area was the ancient Greek and Roman port of Knidos, where even today the ruins of two theaters, homes, and Roman baths still exist. It was here the first statue of a naked woman was created by a famous Greek sculptor. Previously, only naked men were immortalized in marble. People came from afar to see the statue of Aphrodite, the goddess of love and good luck for sailors.

Although the statue is now in a museum, I pictured her standing here in the temple at Knidos. It's been said there was a rear door to the temple allowing sailors to admire and touch the statue before departing on a long voyage, thereby bringing good luck. The harbor itself is a beautiful one with surrounding hills and ruins that lend themselves to picturing life here thousands of years ago. Much of Knidos is undiscovered and will require extensive excavation.

Continuing west, we visited the towns of Gokagac and then Datcha. The city of Datcha resembles a small scale Bodrum but with fewer tourists. There's a street bazaar here with a tent canopy covering of the streets. On our return, we didn't stop at Knidos a second time due to high winds and the lack of protection but instead stopped at the small resort area of Kalaboshi. We went ashore to enjoy dinner at a small restaurant located along the beach. The meals were inexpensive, and the owner spoke fluent English and made a special effort to make us feel welcome.

The following morning we left to return to Turgetreis and D-Marin to prepare for our departure for South Africa in a few days. As we left the protection of the leeward shore, we encountered 38-knot winds with higher gusts. Putting two reefs in the main, we continued sailing for another three hours until reaching our destination, D—Marin.

After returning, *Paroo* sent their laundry to the marina for washing and folding. Because we had dirty clothes they invited us to include ours. Unfortunately, we included a new red blouse that resulted in turning all their clothes pink. Each day, I saw David he appeared to wear the same clothing, everything he owned was a pretty pink. The next several days before our departure to South Africa we utilized our time to make last-minute arrangements and finalize projects. Instead of staying aboard *Tahlequah* on the hard, we elected to stay at a room in the Yacht Club making our transition back to land easier.

Relocating from a boat to land is not an easy transition and often requires adjustment. The movement of a yacht, the splashing of water on the hull, the security of a small "V" birth, the familiarity of restricted space, and the loss of the daily observance of the sun descending into the horizon all lend themselves to a feeling of disorientation. Human nature facilitates the adjustment to make the best of all situations.

Knowing we would be away several months, we arranged for several projects to be completed during our absence. Minor osmosis (water) was discovered in the hull, which is normal for the age of the boat. We arranged to have the hull stripped, dried, resealed and bottom painted, a process taking six months. New batteries would replace the old ones, and a modification of the stern rail made to facilitate the addition of a walkway, making it easier to board and leave *Tahlequah*.

We installed new engine mounts, propeller shaft, rebuilt the turbocharger, and replaced all running rigging before leaving. Following a shake-down cruise with mechanics aboard, everyone was satisfied *Tahlequah* was in excellent mechanical condition and ready for another season of sailing in the Med. Our future plan (winter 2006) included the replacement of *Tahlequah*'s fixed rigging, repainting of both masts, replacing internal pulleys and adding new radar capable of tracking 10 targets. To our thinking, this made *Tahlequah* a new yacht and ready for a future Atlantic crossing.

In preparation for our trip to South Africa, Helen and I went to town in search of a hair salon. Helen being the braver of us volunteered to be first. My turn came, knowing the barber spoke little English, I emphasized *not short please*, three times. The barber repeated *short* three times. Discretion being the better part of valor, I attempted a hasty exit only to be beaten to the door by the barber who through sign language requested I wait until he returned. Moments later, he returned from a clinic across the street with a medical doctor dressed in whites. As the doctor spoke fluent English, he translated my hair grooming wishes to the barber. Although unusual, all went well, and both Helen and I were satisfied with the results.

We looked forward to visiting our close friends from the Rally, Ivor and Bernice. Ivor and I accepted an invitation from a Swiss-based International Surgeons Group to attend their semi-annual conference as guest speakers held this year in Durban, South Africa. I'd be talking about our experience in the tsunami in an ongoing effort to raise funds for the Coco De Mer Tsunami Appeal. This is a Tsunami Appeal created and supervised by Yachtsmen (Barry and Christiane Cager).

Following the conference, we have been invited by Ivor and Bernice to tour South Africa, followed by two weeks in the bush at their Safari Preserve. I plan to shoot as many animals as possible with my laser-guided digital Minolta.

CHAPTER 24

We left Turkey on August 18, and flew to South Africa, arriving in Johannesburg 28 hours later. Our friends Ivor and Bernice frequently talked about their love of South Africa and in particular their love of the bush. Since knowing them, they promised one day to share the best parts of Africa with us.

En route to Heidelberg from the airport, we stopped along the way to pick up suitable clothing for the bush. For the next three days, we stayed in Ivor's home, walked in the gardens, played with his dogs, and toured his feedlot. Ivor showed us our itinerary for the next three weeks; we were thrilled and looked forward to the opportunity to see so much of Africa.

Leaving Heidelberg, we packed the Land Rover and drove six hours to Simunye, Zululand. Arriving, we were given the choice of riding horses or driving in the Land Rover to the valley below to visit a Zulu Village. Due to my hip (remember, they call me "gimpy"), Ivor volunteered to take us by Land Rover. At dark two, Zulus bearing lighted torches came to guide us back to their village. Ivor spoke some Zulu, contributing to our welcome and success with the tribesmen. We also had a guide as a translator who helped explain village customs and traditions.

After walking across a wooden bridge, we reached the village entrance. Our guide requested permission to enter the village. The chief dressed in a leopard skin came to the gate and welcomed us into his village. It was explained that this chief was the grandson of the great Zulu warrior who fought the British at Blood River. Although they fought bravely, the Zulus lost thousands of warriors due to the firepower of the British. The moment lives on today and is remembered as one of the saddest moments in South African history.

Sitting around the fire were several young Zulu men dressed in traditional clothing consisting of animal skins and one of the chief's two wives who prepared homemade beer for us made from sorgum. Needless to mention, the chief's wife drank first lest it be poisoned. (Well, of course, I'd do the same for Helen).

Following dinner, we were escorted back to our huts by two torchbearers. Back at camp, we were seated around a fire and ate traditional foods prepared by the village women. Following dinner, we heard the beautiful singing voices of villagers crossing the bridge led by torchbearers. Two Zulus rubbed water on drum skins and heated them by the fire to tighten them for improved sound. These drums were played as the young village men and women approached us.

All wearing traditional clothing, the men wore animal skins and carried spears and shields. The women wore beautiful, colorful, beaded skirts and were bare-chested. It's the custom of young Zulu women to be bare-chested, covering themselves only after marriage. Zulu women select their own husbands and have the right to turn down any suitors. Once a Zulu woman selects a husband; it's the prospective husband's responsibility to provide her family with cattle.

The men are strong and danced to the sound of the beating drums. Each young man walked forward, quickly raising his foot to the level of his head and stamped on the ground. Their strength causes the ground to rumble as they stamp their feet. This is performed to a hastened rhythm that continues to become faster and faster. It's an impressive sight and clearly intended to intimidate an enemy. If this isn't enough the four young girls repeat this display of prowess and are no less intimidating. The strength of these handsome people is remarkable as both men and women are solid muscle. During this entire performance, I took still pictures. Ok, so I took more pictures of bare-chested women than men, but I've mastered this subtle practice since leaving the South Pacific.

Retiring to our huts, Helen and I enjoyed a romantic evening by candlelight. Following breakfast, we returned once again to the Zulu village, where our guide explained village culture, and through demonstration, we learned how spears were made and war strategies developed through the millennium. We were invited to throw spears, but thankfully my survival didn't depend upon accuracy or we wouldn't have made it through the day. In parts of Africa today, the lion is hunted by spear. Our guide explained that the women do all the work, and the men are strong and brave (no, he didn't smile).

We were introduced to the Noisome, a medicine man. He danced to the sound of drums, chanting as each of us was invited to breathe burning henna plant from a gourd. The smell was intoxicating. We watched the chief's second wife grind mace (called mille pop), popular rice-like food we've frequently enjoyed in South Africa. Half-a-dozen cattle fenced in the center of the village belong to the Chief and are sold to raise money when the Chief deems it necessary. This can be compared to a living bank account. Following the purchase of beadwork and making a customary contribution, we left the village by four-wheel drive.

Continuing to drive through Zulu Land and hundreds of hut villages along the countryside, we were now en route to Durban, South Africa. The city of Durban is wonderfully modern with all the amenities and infrastructure of a US city. Ivor and I were interviewed by a local Durban newspaper. In an open session, we described our experiences during the tsunami in Thailand. Ivor talked about the effects of the tsunami on the round-the-world Blue Water Rally, and I spoke on the tsunami at Phi Phi Don Island, Ground Zero. I gave a 40-minute presentation that went beyond my allotted time and had to be reminded politely to hurry.

The following morning at breakfast a woman came to our table to say her husband (a doctor) hadn't slept all night, having nightmares about the tsunami. Talking about the tsunami is emotionally draining even nine months after the event. I have agreed to do one more presentation for a local Yacht Club in the US for the purpose of continuing to raise funds for the Coco De Mer Tsunami Appeal Fund. In the future, I will limit myself as we need to move on with our lives.

On the fourth day in Durban, we drove to the airport where Ivor's plane was landing to transport us to Mozambique and the Marlin Lodge located offshore. Here we met Ivor's pilot, Dion, a charming chap and a welcome addition to our adventure holiday. Each time he landed, we gave him a round of applause for his success in getting us back on the ground. En route to Marlin Lodge, we stopped to check on the progress of Ivor's new home under construction in Mozambique. From the airport, we passed many grass huts and children walking to and from school, and many smiled and waved. It isn't uncommon for children to walk one to two hours each way to school.

Marlin Lodge is owned by a close friend of Ivor and Bernice who sometimes share access to each others bush camp and lodge. While we were at the Marlin Lodge, its owner was enjoying several days at Ivor's Bush House and Camp bordering the Kruger National Park. Tours of the island in a Land Rover, a fishing excursion, meals, including wonderful dinners in the lodge, and even Helen's facial were all included. I jokingly said to Ivor when they came to our home in Venezuela, we would reciprocate with an evening of bowling to make up for the wonderful holiday in South Africa; fair is fair.

Although the island is a national park, the lodge was constructed on leased property through the government to encourage tourism in the park. Since the Portuguese gave independence to Mozambique during the 1960s they worked hard at encouraging tourism and sources of outside investment to the area.

Following four wonderful days at Marlin Lodge, we left aboard Ivor's plane for Livingstone, Zambia, for the purpose of seeing the wonder of the Victoria Falls, and a wonder it truly was. Checking into the Royal Livingstone, we prepared to have an exciting time in Zambia. We took advantage of high tea, something Helen and I have grown accustomed to every afternoon since joining the British Blue Water Rally.

Following check-in Ivor, Bernice, Helen and I hiked alongside the Zambezi River, one of the five largest rivers in Africa and on to Victoria Falls. Victoria Falls is 200 meters high. We were there during the driest part of the season making it possible to see below without the usual rising mist. It was a spectacular view and is one of the natural wonders of the world.

En route we encountered a family of baboons. The male was very large and was intent on baring his teeth and making aggressive motions toward Helen. (I may have done that myself when first meeting Helen). I lunged forward showing him my hiking stick with titanium spike.

He stopped, sat down staring at me as if to say, *Go ahead, make my day.*

I shook my stick at him, and he rose up and sauntered away unimpressed with my display. Knowing these animals were not only aggressive but potentially hostile, I was prepared to take a good swing if necessary. However, we weren't worth the effort, and he looked for other easy pickings.

Because the panorama of the falls is beyond anything we've seen, we took many still pictures and videos throughout the afternoon as we continued walking alongside the falls. Those seeking adventure white water raft at the bottom of the falls and must be airlifted out by helicopter at the end of the day. At a nearby bridge crossing the gorge, people bungee jump several hundred feet. That is a sport that yours truly prefers to watch on television (ok, ok, so I blamed it on my hip).

During dinner, we watched the day's events on one of several monitors showing white water rafting. One raft transporting 20 Germans was swamped and turned over during the worst of the rapids. Oh well, something to tell their grandchildren about.

Zebras and hundreds of monkeys roamed freely about the grounds, having free access due to the hotel being located in the African bush. One morning while sitting on the patio a gardener asked if I had seen the giraffes nearby. He offered to take Helen and me across the path and into the bush where we observed and photographed two young giraffes.

At 1600 hours we boarded the African Queen and sailed the Zambezi river for the purpose of seeing wildlife and appreciating a wonderful sunset. We saw wild elephants, hippos, and fish eagles soaring overhead. The African sunset was spectacular and brought our first day in Zambia to a wonderful end. All you could eat and drink contributed to the success of the day.

During the next three days in Zambia, we continued to enjoy walks in the bush and went to a nearby market where many carvings were for sale. We purchased an African board game we had seen people playing with earlier in the day. The obvious benefit of having our digital camera purchased in Australia is that we continued to take hundreds of pictures. Back home we have many boxes consisting of thousands of pictures we no longer look through. I reassure myself it's our intention to locate a few pictures taken during our circumnavigation worthy of framing.

CHAPTER 25

One hour forty minutes later, we landed in St. Petersburg, completed South Africa entry paperwork and were once again airborne for Timbavati, Ivor's Bush Camp adjacent to the Kruger National Park. We spotted the airstrip and immediately descended to touchdown minutes later.

Ivor's Land Rover with staff met us at the airstrip. Our Land Rover was the open type with one external seat positioned forward of the hood for our tracker whose purpose for the next several days was to help locate large game for us to photograph. Because of my hip, I sat in front next to Ivor's game warden behind an elephant gun mounted on the dashboard forward of me. I contemplated if attacked I'd have to shoot myself.

Twenty minutes later, we entered the camp through an electric fence surrounding their home. It was a beautiful, functional, and practical home with traditional African thatched roof. A large fireplace surrounded by sofas and chairs accented the social area. The rear of the house had an immense covered patio area with several sofas, dining table, and lounges. A step-down pool with surrounding boulders complemented the patio. Beyond the rear electric fence was a watering hole where animals could be observed from the rear porch.

Storing our backpacks, we immediately dressed in kakis and boarded the Land Rover. Ivor's Game Warden, Rob, drove while Martin tracked from his personal seat located on the front hood of the Rover. Armed with photographic equipment, Ivor, Helen, and I were intent upon getting good pictures during the next four days. To our surprise and delight, we stumbled across two male lions lying on the side of a dirt road. By the appearance of their distended stomachs, they had consumed a seven-course meal. Undisturbed by our presence they continued going about their rest.

Our guide Rob explained that large game here were conditioned to open Land Rovers and did not distinguish between the vehicle and ourselves. He alerted us not to put extremities beyond the profile of the Rover, stand up, or attempt leaving the vehicle, as we would become fair and easy prey for our magnificent friends. Advice we accepted without question.

It was explained to us two male lions frequently pair up to share a pride (female lionesses) for two or more years or until another lion challenges and replaces them to dominate the pride. We were told these lions dominated this pride for five years and would likely soon be replaced. Continuing on further we came across a leopard and her four-month-old cub. The leopard ignored us in the same manner as the lion; we took photos of her until dark. A leopard is both a ferocious and magnificent-looking animal and one you wouldn't wish to encounter on foot. It was clear to us these animals have few predators in the wild.

The first night in the bush, we were awakened by the roar of lions heading south in the Timbavati. Each morning we awoke at 0530 to begin our trek in search of game at 0600. At 1030 we returned to Bush Camp for breakfast. At 1500 we began a second trek and completed the day's activities by stopping at Ivor's second Bush Camp, U'Bali. It's a magnificent location with a farmhouse located atop a hill overlooking a watering hole and range below. Each evening Rob and Martin set up a portable table, opened a bottle of wine while we watched the sunset enjoying sundowners. No site on earth can be more beautiful, and it reminded us of the movie *Out of Africa*.

On the second day in search of elephant herds, our game warden turned and asked Ivor if he could smell elephants. Apparently, Ivor has a gift for smelling and locating elephant herds when no one else can find them.

Turning his nose to smell the wind, Ivor said, "I think that way," pointing off to the left. Several hundred paces down the road Ivor again said, "The smell is stronger here. Turn left now, and a few more paces that way, through those trees, there."

Sure enough, Ivor had led us to a herd of elephants.

I thought to myself, *Ivor can always get a job-tracking elephants with his nose. Wish I had some type of talent, can't even play a musical instrument.*

African elephants are much larger than those we rode in Thailand or saw at US zoos. Rob explained elephants are a matriarchal society, and when one female elephant becomes too old to lead, she will transfer responsibility to her sister or another elephant. We enjoyed seeing a young elephant (two months old) follow her mother foraging through the growth. Seeing elephants over the next several days, it became apparent to us how destructive they can be. They defoliate and topple trees throughout an entire area. When the population grows beyond what an area can support, the South African government will tranquilize and transport the elephants by helicopter to another area—a costly solution.

Because it was the end of winter and green foliage was becoming sparse, elephants walked through Ivor's electric fencing with ease to eat his remaining and protected vegetation. It's a losing battle attempting to stop an elephant; chasing them doesn't result in solving the problem, as they return after dark. During the remainder of our days at Timbavati, we saw the Big Five: elephants, lions, hippos, rhinos, and Cape buffalo.

I was surprised to learn that locals feared the Cape Buffalo more than any other animal, including the lion. The Cape (Southern) Buffalo is the only animal that will lay a trap for its pursuer. It can take up to ten rounds to stop a Cape Buffalo. It will run off, intentionally leaving misleading tracks, circle back, and wait in hiding for its tracker. Seeing its prey, it will attack with deadly ferocity. No other animal will attack its prey in this manner.

One evening, we came across two lions devouring a Cape Buffalo carcass. They worked together tearing the sinuous muscle and devouring what was left of the animal, including its bones. We parked less than two meters from the lions and remained close by, taking pictures and flashing away until dark. It amazed us how animals seem unbothered by the flash or noise of cameras.

Knowing the speed and strength required to bring down a Cape Buffalo, we could appreciate what little effort would be involved in killing a human. At one time, it was believed only old and sickly lions attacked humans. New studies indicate that 60% of lions that attack humans are healthy. Watching lions devour their prey at close proximity helped us appreciate how vulnerable we are in the food chain.

On the third day in the bush, we returned to U'Bali for sundowners. Sitting there appreciating the sunset, we saw an impala walking across the path nearby. Minutes later, we saw a leopard stalking the impala. As the leopard came into the open, it saw us and sat down to observe us. Later it stood and began slowly moving toward us. We sat in our chairs mute and motionless so as not to alarm the leopard. Knowing the rifle was back in the Land Rover, I searched about for means of defense. Helen and I were seated in a double chair, too heavy to raise in defense effectively. Knowing running from an animal insured I would be torn apart, I decided if the animal showed any signs of attacking, I would run towards him yelling and screaming. As the animal got within five meters of us she began to walk past the camp slowly.

I learned later that each of us had developed our own plans for defending ourselves if the leopard attacked. Even our Game Warden Rob was relieved when the animal left. His words were "that was an incredible experience; you won't see that often." (Yes, he looked relieved.)

Our final full day of trekking, we returned from U'Bali cautiously in total darkness. Our tracker now armed with a mega spotlight guided us back slowly. In darkness, it was impossible to know where we were.

Suddenly, we came upon a campfire, but instead of passing by, Ivor said, "What's this? Must be neighbors. Come on let's stop and introduce ourselves."

I'm thinking, *Boy, this guy has nerve, stopping uninvited on someone's barbeque.*

Ivor jumped out of the Land Rover and said, "Hi, neighbors, want you to meet good friends of ours."

I got out and walked up to an attractive young lady saying, "Hi, I'm Ed; this is my wife Helen," all the time thinking to myself, I know this woman.

After falling lock, stock, and barrel for this one, I discovered these people all worked for Ivor, and the woman was our game warden's wife, whom I met back at bush camp. Oh, well, Ivor got me again. Ivor had arranged a special brie on our final evening in the bush.

On the fourth and final day before departing, we said goodbye to Rob, his wife, Yevette, and our tracker, Martin. This had truly been an extraordinary experience and one we will forever remember. We began our six-hour drive back to Heidelberg via the longer mountain route to enjoy the beautiful mountainous landscape. Stopping along the way, we purchased African bowls and carvings to bring back to the US for family and friends. We arrived back in Heidelberg for a few days of rest before taking another flight to Cape Town.

Cape Town is a wonderful tourist location with beautiful coastline, attractive city and seaport, and beautiful views from atop Table Mountain. During these three days, we saw all of these things and more. In Cape Town, we stayed at the Victoria and Albert Hotel located on the waterfront. The area was abundant with tourists enjoying sightseeing boats, beautiful malls, arts and craft shops, and delightful restaurants for all tastes. Because the hotel was renovating their third floor, hard hats were left in each room making light of the situation. One morning I awoke at 2:00 to use the head and returned wearing my hardhat and nothing else. Waking Helen, I did one of several athletic poses to be greeted by hysterical laughter.

"Forgot your tool belt there, fella," Helen managed to get out.

Taking a cable car to the top of Table Mountain, we were treated to a spectacular panoramic view of both the Indian and Atlantic Oceans. Walking around the edge, we came across a young man climbing up the cliff and over the protective railing. Apparently, it's possible to parachute halfway down the mountain then climb back up. South Africa is clearly not a litigious society by any standards. The coastline offers spectacular views reminiscent of the California coastline. We saw a penguin nesting area and had lunch nearby before taking a half hour boat ride to seal rock. Yes, there were seals.

Driving to the Cape of Good Hope was the highlight of the afternoon. We felt we accomplished another great milestone in our travels and asked Ivor to take our picture in front of the sign as proof of this accomplishment. Seeing two sailboats sailing along the coastline, I wished I was aboard *Tahlequah* at that very moment. Although this area is infamous for bad weather, we were fortunate to be here during a perfect weather period.

We took a second cable car ride to the hill above the Cape to the lighthouse at the top. A baboon grabbed an unfortunate fellow's bag filled with apples, dumped it upside down, rolled the bag up, and threw it back at its owner. The baboon was cooperative enough for Helen to stand nearby while I took their picture together.

The following morning, it was back to the airport for the return flight to Johannesburg. Then we were off to Heidelberg where we spent the remainder of our time with Ivor, Bernice, and their sons, Justin, Avron, Matthew, and families. Ivor requested I talk to his Senior Staff about the elements of World Class Manufacturing. I spent several hours putting together a presentation and spoke with his staff at the Heildelberg office. Although the group was larger than anticipated, I enjoyed the challenge, and it reminded me of when I worked for a living.

The next two days, Ivor arranged for his staff to take us on local tours in Johannesburg and Pretoria. We visited a gold mine descending 200 meters. Although the mine originally descended two miles beneath the surface, we were prevented from going deeper due to flooding. The tour was a fantastic one and provided us a genuine appreciation of what mining was like. I can't imagine working at those depths and in those conditions day after day. The Crown mine was the largest mine in Africa.

Next we went to the Apartheid museum, a moving and stirring experience. The museum was constructed to be prison-like. Entering, we were requested by sign to enter the Whites Only side of the prison. Once inside, we passed long passages with pictures of identity cards describing one as white, colored, or black.

The first large exhibit we saw was one of US slavery and racism in the South. As an American, that was an important exhibit, and it brought the issue of racism home and not just something that a far-away country had to deal with. We saw pictures of hangings, shackles, videotapes of Southern politicians proclaiming segregation, and Ku Klux Klan hooded manikins.

Growing up in Arizona in the 1950s, I experienced schools as they were becoming integrated. The first black student was placed in our class. Before he arrived, the teacher made a special point how important it was to make him feel welcome. I invited Jake to come home with me three times in one week.

One afternoon, my mother took me aside and said, "You need to have white friends."

My mother then proceeded to her bedroom, closed the door, and began sobbing. Her tears had a greater effect upon me than her words. Although my mother was likely responding to what the neighbors thought, I knew her true feelings. My mother's tears helped remove racism from my heart.

Racial prejudice is a sad part of American history, and it continues today in different forms. It's easy to point a finger at South Africa.

Continuing to explore the museum, we saw the history of Apartheid in South Africa through the present day. The Apartheid museum stressed unity between all peoples.

The words of Nelson Mandela inscribed on the side of the building said it all, "What good is it to cast off chains if we can't respect each other's differences?"

The following day we went to Pretoria to visit the Voortrekker monument built in 1938 to commemorate the trek northward of the early Dutch pioneers. The monument's highlight is a stream of sunlight penetrating the dome to cast itself upon a symbolic marble box once a year on December 16. Each year, people come to attend a special church service in observance of their forefathers, who settled these lands and defended them from the British.

Following a tearful farewell, we left South Africa on September 15 to fly to the US. Although I'd managed to limp around South Africa using a hiking stick, it was clear to everyone that without a hip replacement, I'd soon be unable to sail or walk.

I looked forward to convalescing, completing our book, and returning to the sailing life. Our future plans included joining the Eastern Med Rally visiting Turkey, Cyprus, Israel, and Egypt. Our long-term plan was an Atlantic crossing and a return to the Caribbean.

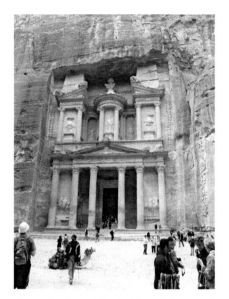

Petra, Jordan—The Valley of the Dead

Petra—Royal Guard Soldier

Petra Desert—Dick York & Ed

Nile, Egypt

Pyramids—Cairo

Luxor, Egypt—children on the Nile

Suez Canal

Ephesis, Turkey

Bernice, Ivor (family), Helen & Ed

Timbavati (Ivor's Bush Camp), South Africa

Cape of Good Hope—Helen & Ed

Helen & Alicemay Wright—our inspiration to Circumnavigate.

Epilogue

One year since the tsunami: Although feeling an emotional attachment to Thailand and a need to return to Phi Phi Don Island for the anniversary of the tsunami, we cannot go this year.

Helen and I continue to receive invitations from yacht clubs and other groups to talk about our experience. Eleven months after the tragedy, strangers come up to us on the street asking if we were the people they saw on *NBC Nightline, Oprah,* or read about in a sailing magazine or newspaper article.

The inevitable question we get is "How did it change your life?"

As often as we're asked this question, it's always a difficult one to answer.

I can say it did change our lives. Helen responds she's now grateful for every moment she's alive and feels a need to make the most of what time she has in life. Things that seemed important to us before no longer are. Our perspective of what's important in our daily lives has been altered dramatically. Helen claims my patience has improved. Because Helen was unconscious much of the time, her experience and mine are different.

Everyday I see that wave approaching, it's as vivid as the day it happened. I still ask myself why we survived when so many died; there will never be answers to this question. I'm haunted by the faces of the people I left behind and will never forget them. People I never knew have become a part of my life. Every December 26, at 10:45 am (Thailand time), I will toll eight bells in memory of those who perished that day.

The second question we're asked most frequently is "What country did you like most?"

We try to explain that each is so different and unique that we've found many places we hope to someday return to. We work hard at explaining it's the people that make a country unique. We emphasize circumnavigating contributed to changing our perception of the world. Visiting a country under sail brings us into contact with local people and their everyday experiences. The open hospitality of Muslims inviting us into their homes was contrary to what we expected as Americans. The people who have the least are those who offered us the most. It mattered not if we were in the South Pacific, Indonesia, or Malaysia, good people can be found everywhere.

As cruisers, we inherit the responsibility of becoming diplomats as we sail around the world. Each of us becomes a spokesperson representing our countries and those who will follow behind us. Leaving good impressions and a clean wake insure others will have the same experience as a welcome guest in a foreign country.

978-0-595-50735-1
0-595-50735-2